T0245219

CAMBRIDGE LIBRARY COLLECTION

Books of enduring scholarly value

Earth Sciences

In the nineteenth century, geology emerged as a distinct academic discipline. It pointed the way towards the theory of evolution, as scientists including Gideon Mantell, Adam Sedgwick, Charles Lyell and Roderick Murchison began to use the evidence of minerals, rock formations and fossils to demonstrate that the earth was older by millions of years than the conventional, Bible-based wisdom had supposed. They argued convincingly that the climate, flora and fauna of the distant past could be deduced from geological evidence. Volcanic activity, the formation of mountains, and the action of glaciers and rivers, tides and ocean currents also became better understood. This series includes landmark publications by pioneers of the modern earth sciences, who advanced the scientific understanding of our planet and the processes by which it is constantly re-shaped.

A Complete Guide to the English Lakes

First published in 1853, this is a comprehensive guide to the British Lake District. It features contributions from William Wordsworth and the geologist Adam Sedgwick, as well as a number of shorter sections by local experts on subjects such as botany and toponymy. The first part comprises detailed descriptions of the major towns and villages of the area, providing recommended routes and excursions for tourists. This is followed by Wordsworth's description of the scenery of the Lake District, offering fascinating observations on the natural formation of the landscape and the influence of human settlement. The latter part consists of a series of five letters on the geological structure of the area, written by Sedgwick between 1842 and 1853. Illustrated with detailed maps of the area, this is a key text for those interested in the history of tourism in the Lake District and its development in the Victorian period.

Cambridge University Press has long been a pioneer in the reissuing of out-of-print titles from its own backlist, producing digital reprints of books that are still sought after by scholars and students but could not be reprinted economically using traditional technology. The Cambridge Library Collection extends this activity to a wider range of books which are still of importance to researchers and professionals, either for the source material they contain, or as landmarks in the history of their academic discipline.

Drawing from the world-renowned collections in the Cambridge University Library, and guided by the advice of experts in each subject area, Cambridge University Press is using state-of-the-art scanning machines in its own Printing House to capture the content of each book selected for inclusion. The files are processed to give a consistently clear, crisp image, and the books finished to the high quality standard for which the Press is recognised around the world. The latest print-on-demand technology ensures that the books will remain available indefinitely, and that orders for single or multiple copies can quickly be supplied.

The Cambridge Library Collection will bring back to life books of enduring scholarly value (including out-of-copyright works originally issued by other publishers) across a wide range of disciplines in the humanities and social sciences and in science and technology.

A Complete Guide to the English Lakes

With Mr. Wordsworth's Description of the Scenery of the Country

Adam Sedgwick
Edited by John Hudson

CAMBRIDGE UNIVERSITY PRESS

Cambridge, New York, Melbourne, Madrid, Cape Town, Singapore,
São Paolo, Delhi, Dubai, Tokyo, Mexico City

Published in the United States of America by Cambridge University Press, New York

www.cambridge.org
Information on this title: www.cambridge.org/9781108017886

This edition first published 1853
This digitally printed version 2010

ISBN 978-1-108-01788-6 Paperback

Drawn & Engraved by W. Banks Edin.r

HEAD OF WINDERMERE FROM LOW WOOD

Published by John Hudson. Kendal.

A COMPLETE

GUIDE TO THE LAKES,

COMPRISING

Minute Directions for the Tourist;

WITH

MR. WORDSWORTH'S

DESCRIPTION OF THE SCENERY OF THE COUNTRY, ETC.:

AND

ON THE

GEOLOGY OF THE LAKE DISTRICT,

BY THE

REV. PROFESSOR SEDGWICK.

Fourth Edition.

EDITED BY THE PUBLISHER.

KENDAL:
PUBLISHED BY JOHN HUDSON.
London:
LONGMAN AND CO., AND WHITTAKER AND CO.
LIVERPOOL; WEBB, CASTLE-ST.—MANCHESTER; SIMMS AND CO.

1853.

ADVERTISEMENT.

Encouraged by the steady sale which has rapidly exhausted the Third Edition of this work, the Publisher has spared no pains to introduce into the present impression such alterations and additions as seemed to him calculated to add to the interest and usefulness of the work.

It has been his aim to combine in the present volume not only an accurate "Guide to the Lakes," in the strict sense of the phrase—a book which will be useful to the Visitor *during* his tour,—but which also, by containing, in addition, subjects of abiding interest, shall be worthy a permanent place in the library.

The distinguished authorship of a considerable portion of the work enables the Publisher to hope that he has gained his object. "*The Introduction,*"—"*the Description of the Lake Scenery*"—and much of the "*Directions and Information for the Tourist,*" are from the pen of the late Mr. Wordsworth, who has left a name now inseparably connected with the district of the English Lakes, the beauties of which he has so admirably illustrated both in prose and "immortal verse."

Professor Sedgwick has kindly furnished another Letter in addition to the Four which have already appeared in former Editions of this Work, bringing up the investigations of this complicated Geological country to the present time. The Appendix also contains a list of all the additional Igneous Dykes which have been discovered, and of the Fossil Organic Remains of the district. The value and interest of any production from the Professor's vigorous and lucid pen need not to be pointed out.

Mr. Gough, of Kendal, (who bears a name well known in the botanical world,) has kindly contributed lists of the rarer plants which the tourist may meet with in his rambles. To the same friend the Publisher is indebted for a copious list of the Land and Fresh-water Shells of the district, which will be interesting to the Naturalist.

The outline Diagrams of the Lake Hills (taken from certain well-known points of view) are from the pencil of Mr. Flintoft, of Keswick, whose accurate knowledge of the Lake District is proved by his beautiful model of the country, which has been the admiration of so many Tourists.

The interesting chapter on the derivation of local names has been supplied by Mr. Nicholson.

For the remaining portions of the volume, original and selected, the Publisher holds himself responsible. The *Tables of Distances* and the *Itineraries* have been carefully tested by personal survey, and compared with those given in Green's Guide to the Lakes,—by far the best and most accurate of the larger works of the kind which have appeared; and, to render this portion of the work still more complete, several new routes and approaches to the Lakes, by railway and steam communication, have been added.

Grateful for the encouragement which this little volume has already received, the Publisher commits the *Fourth Edition* of the Work to the continued liberality of the Public.

Kendal, July, 1853.

CONTENTS.

DESCRIPTION OF THE SCENERY OF THE LAKES.

SECTION FIRST.

VIEW OF THE COUNTRY AS FORMED BY NATURE.

SECTION SECOND.

ASPECT OF THE COUNTRY AS AFFECTED BY ITS INHABITANTS.

SECTION THIRD.

CHANGES, AND RULES OF TASTE FOR PREVENTING THEIR BAD EFFECTS.

SECTION FOURTH.

STAGES.

	Miles.
Lancaster to Kendal, by Burton	22
Lancaster to Kendal, by Milnthorpe	21
Lancaster to Ulverston, by Levens Bridge	35½
Ulverston to Hawkshead, by Coniston Water Head	19
Ulverston to Newby Bridge*	8
Hawkshead to Ambleside	5
Hawkshead to Bowness	6
Kendal to Birthwaite (Railway terminus)	9
Kendal to Ambleside, ...	14
Kendal to Ambleside, by Bowness	15
Kendal to Patterdale (Ullswater), by Ambleside	24
Kendal to Patterdale, by a new and pleasant road through Troutbeck, which leaves the Ambleside road on the right, a short distance beyond Ings Chapel	18
From Ambleside round the two Langdales and back again	18
Ambleside to Ullswater	10
Ambleside to Keswick	16½
Keswick to Borrowdale, and round the Lake	12
Keswick to Borrowdale and Buttermere	23
Keswick to Wasdale and Calder Bridge	27
Calder Bridge to Buttermere and Keswick	29
Keswick, round Bassenthwaite Lake	18
Keswick to Patterdale, Pooley Bridge, and Penrith	38
Keswick to Pooley Bridge and Penrith	24
Keswick to Penrith	17½
Whitehaven to Keswick	27
Workington to Keswick	21
Penrith to Hawes Water	27
Carlisle to Penrith	18
Penrith to Kendal	27

* Steam-boats ply from Newby Bridge to Bowness and Ambleside, up the Lake of Windermere, two or three times a day, during the summer, Fares very reasonable, and accommodation on board excellent.

A GLOSSARY,

ETYMOLOGICAL AND EXPLANATORY, OF THE NAMES OF HILLS, LAKES, RIVERS, ETC., OCCURRING IN THIS VOLUME.

Mountainous districts, generally speaking, have been so many refuges for the primitive dialects. The reason of this, even if our space permitted, it is hardly necessary to explain, as it must be apparent to every reader of history and every reflective mind. Hence we find in the *names* of the striking natural objects of this district, so many descriptive epithets signifying the same thing, and all proceeding from the oral dialects of the early inhabitants. We recognize, as applied to the names of the Lake Mountains, no fewer than twenty-four different words in the Celtic, Saxon, and Teutonic tongues, each signifying *hill:* and there appears to be almost an equal number expressive of *water:* hence it is that in the composition of many of the names there are so many repetitions (triplications in some instances) of the same meaning. Most of the names that still pertain to the hills, lakes, &c., come from the Saxon, Dano-Saxon, and Teutonic dialects, and it is natural to suppose that after the invasions by the Saxons and Danes, these names have replaced, and been made to obliterate, the *previous names* in the Celtic or British tongue. Such words as Glaramara, and Blencathra (now Saddleback), look like remains of the pure British, but we have no good clue to their signification.

Some of these names, as it will be seen, have their origin in the *external appearances* or *configuration* of the object: this class refers chiefly to hills. Others are derived from some *essential quality* or *peculiarity* of the place or thing designated: these have reference mostly to lakes and rivers. Again, others are so denominated from the fact of wild animals having abounded there, as the wild boar, deer, goat, cat, &c. The names terminating with *thwaite*, as Legberthwaite, Tilberthwaite, &c., have evidently received their appellations on the introduction of agriculture.

GENERAL TERMS.

BARROW (Ang.-Sax. 'beorgh') a hill, natural or artificial.

BECK (Sax. and 'bek,' Dan.) a small stream or rivulet.

CAM, COMB (Sax.) properly the *crest* of a hill, as the comb of a cock.

COOM, OR COVE (British 'cwm') a valley, opening between hills.

DAL, (Danish) dale, a little valley.

DEN, a dale or glen.—DON, or DUN, a smaller hill.

DOD, applied to a smaller hill to distinguish it from a greater; for example—Skiddaw Dod.

DORE, (British 'dwr') water; a word that enters largely into the composition of names in the Lake district. *Dore* is applied also to an opening between rocks.

Force, a waterfall.
Gate, ('geat,' Sax.) a way.
Garth, an inclosed piece of ground.
Ghyll, (Isl.) a fissure in a mountain, or between two mountains.
Grange, a farm or house near water, but the Farm of a Monastery or Baronial establishment was called the Grange.
Hag, a general term used for an inclosure.
Hawse. (Sax. 'hals') a throat, or gullet.
Hirst and Hurst (Sax.) a wood, or grove.
Holm, a piece of land, either surrounded by water, or washed by one or more streams—either an island or a peninsula.
How, (Teut.) a small hill. Chaucer uses *how* for a cap or hood.
Hul, (Sax.) a hill.
Ings, low meadows.
Keld, a well.
Knot (Sax.) applied to hills with a marked prominence or protuberance in the same sense as a 'knot' on a tree.
Man. A factitious eminence set upon a hill. *Maen* (Brit. ?) is an old word for stone, however, and the 'man' of the mountains is always of stone.
Nab, (Sax. 'cnep') the 'neb' or nose of a hill. The bill of a bird is called its 'neb.' Thus the Highlanders prayer—

" Fra' witches and warlocks and lang-*nebbit* things," &c.

Ness, Neese, or Naze ('nese' Sax.) a point of land projecting into the water. Thus, Bow*ness*. *The Naze*, in Norway, and on the coast of Essex.
Pen (Brit.) hill—whence, also, 'Ben,' B and P being convertible.
Pike. ('pec' Sax.) peak.
Raise. A tumulus formed of heaped-up stones.
Scar or Scaur (Su.-Goth.) a steep escarpment of rock.
Slack (Su.-Goth) 'slak') a depression in the summit line of a hill, or, generally, *a hollow*.
Syke. A rivulet.
Tarn. A small mountain lake.
Thwaite. A piece of land cleared from wood.
Wyke. A bay or creek.

NAMES OF PLACES.

Ambleside (p. 39). As this name was formerly spelt *Hamelside*, and is still pronounced by the vulgar, *Hamelsed* or *Amelsed*, it may be derived from *Ea* or or *Eau* (water), *mel* (a brow). Water from the sides of the brows.
Applethwaite (p. 44). *Ea-pul-thwaite*. The two first syllables are the reduplication of *water*.
Bassenthwaite Water (p. 96). *Bass* is still the provincial term applied to the fresh-water fish, the perch; in this sense,—'water abounding with bassen,' (plural of perch).
Borrowdale (p. 73). *Boar-dale*, or *Borough-dale*. Perhaps a literal variation of Barrow-dale.
Bowfell (p. 84). A bowed, or arched hill. Very applicable.
Bowness (p. 25). A round-pointed promontory. Sometimes written *Bullness*, which has the same meaning and derivation.
Brathay (p. 39). Water from the *brae?*
Brotherilkeld (p. 64). *Broad-dur-ail-keld*—a broad water from the keld or spring.
Buttermere (p. 87). *Bode-tor-mere* (Sax.) or *Booth-tor-mere*—the lake of a village by the hill.
Carl Lofts (p. 108). (Brit. 'caer;' Sax. 'loft')—a high dwelling.
The Carrs. Probably the *Scars*.
Catchedecam (p. 102). Probably the high-*crested*, or high-topped hill where wild *cats* abounded. The old spelling was *Cats-sty-cam*. Sty, a ladder. In Westmorland we still call it a *stee*.
Cat Bells (p. 75). *Bael* (Ang.-Sax.) is a signal fire, or beacon; but there is no record of this mountain having ever been one of the beacon hills.
Causey Pike (p. 86). Causeway Pike.
Cockley Beck (p. 63). A winding or rugged stream.

CONISTON (p. 12). A town (*ton*), at the head (*con*) of the lake (*is*) Brit. Some take it to be a corruption of *Konygs-ton* or *King's-town*.

DERWENT (p. 70). *Dwr-gwynt* (Brit.) the windy lake. This lake is remarkable for gusts of wind. Or, *Dwr-gwyn* (clear) water.

DONNERDALE (p. 14). Some think the first syllable a contraction of 'Duddon.'

DUDDON (p. 12). *Dod-den*, the lower, lesser, or inferior valley.

EASEDALE (p. 59). *Eas*, or *Is-dale*, water dale.

ELTERWATER (p. 46). *Ael* (Brit.) *great* and *Tor* (Sax.) hill. Water from the great hill, or the water beset with elders or alders.

ESKDALE (p. 63). Esk, and Ask, mean *Newt* or *Lizard*. Both words also signify water: and the latter is the more probable derivative.

FAIRFIELD (p. 46). *Faar-feld* (Danish). Sheep pasture.

GATESGARTH OR 'GATESCARTH' (p. 67). A gate or road over the Scar—which is the case in this instance. Or, 'gate' may in this place be a variation of 'Goat, from the wild goats, '*Goat-ca*' is the name of a hill near Gatescarth.

GLENCOIN (p. 99). *Cyna* (in the Saxon) is a 'cleft' or 'fissure," so Glencoin is a reduplication of the same word. BURN says, from 'cuna' (Fr.) (quain)—a corner. If the latter derivation is preferred, the name has been adopted since the Norman Conquest. Antiquaries will prefer the former.

GLENDERATERRA (p. 86). A *glen* conducting (*dwr*) water from *turret* the hill or eminence.

GRASMERE AND GRASSMOOR (pp. 58, 86). Formerly spelled *Gersmere*. From *Gres* (Sax.) 'grass." The lake of grassy banks. Or, *Grismere*—the lake of the wild boar.

GREENUP (p. 77). A verdant upper or higher plot of ground. *Up* is a common adjunct in contradistinction to *lor* or *lower*. Instance, 'Upton,' 'Lorton,' &c. &c. Or, *Gren Hope*: Hope is often corrupted into *up*, *op*, or *ip*. Very common in Northumberland.

GRETA (p. 71). River. Dr. Whittaker supposes this river to take its name from the 'greeting,' or weeping tones of the water down its channel.

GRISEDALE PIKE (p. 86). From 'Gris,' wild swine.

HAMMAR SCAR (p. 59). *Hamur* (Sax.) enters into the composition of the names of many of the Scandinavian hills.

HARRISON STICKLE (p. 41). *Stigle* (Sax.) an acute point. Harrison is evidently a personal name used to distinguish one of 'the Pikes' from the other. Hence our word 'stile,' and 'steel.' 'Steel Pike' was the ancient name of this hill, as Mr. WEST has it—and we would like to see this name restored.

HARTSOP (p. 100). *Harts-up* (so pronounced). The hill of the red deer.

> Where stalked the huge deer to his shaggy lair.
>
> WORDSWORTH.

HELVELLYN (p. 101). Hel (*hill*): gwal (*wall*); lyn (*lake*)—a hill that forms a wall or defence for the lake. Some derive Hel-*bel*-lyn from 'Bel' or Belinus, the God to whom sacrificial fires were lighted upon hills.

HINDSCARTH (p. 87). The Shepherd's hill.

KESWICK (p. 68). We can make nothing of the pre-fix of this name, unless it may be deemed an abbreviation of *Caester* (Sax.) a fortification.* We incline to this hypothesis, and think it the same word as '*Kearstwick*.' Or Kesh is the provincial name still used for the water-hemlock—*Kesh*wick, the village by the keshes.

KIRKSTONE (p. 48). Some derive the name from the rock at the top of the hill.† But there was both a Cairn and a Druidical Altar near the summit of this hill, and we would rather refer the name to that Altar.

LAMPLUGH (p. 92). This name has reference only to the soil of the place. *Lam* (Sax.) is loam or clay. The last syllable explains itself. Mr. NICOLSON (*History of Cumberland*) says 'Glan-flough' (Irish) *dale-wet*.

LANGDALE (p. 40). or *Langden*. Long valley.

LANGSTRETH (p. 62) Long street or way, from 'stret,' (Sax.)

LEGBERTHWAITE (p 61). *Leigh* (Sax.) a meadow, whence ley, *bera* (Sax.) barley; *thwaite*, inclosure. An inclosed barley field.

* See Mr. WEST's Guide. p, 149.

† This block—and yon whose church-like form
Gives to this savage pass its name.

WORDSWORTH.

LINGMELL (p. 66). A *brow* (mæl, Sax.) remarkable for 'ling,' or heather.

LODORE (p 73) and LOWTHER (p. 106), are evidently the same; *Lodwr*. In a very old book, Lowther is interpreted *black-water*.

LYULPH'S TOWER. (p. 49). Some suppose from Lyulph, the first Baron of Greystoke.

MATTERDALE (p. 99). *Mater*, or mother dale, if Pater-dale be adopted. The two dales are adjacent. See a Note in *Nicolson's History of Cumberland*, p. 367, where it is described as Materdale.

MICKLEDORE (p, 66). Greater door or opening. This word 'dore' is sometimes applied to the mouth of a pass.

NANBIELD (p.22). *Nant* (Welch) a fall; and *bield* a sheltered place The spot to which this name applies is a pass crossing from Kentmere to Mardale.

PATTERDALE (p. 48). Perhaps *Pater* or father dale. Some say *Patrick-dale*, from the Patron Saint of Ireland.

PENRITH (p. 104). Pen-*rhydd* (Celtic) red hill. In Wales rhydd is still pronounced *rith*.

PIKE OF STICKLE (p 42). *Pike*, a *peak* (Sax.) *Stigle*. See Harrison Stickle.

PORTINSCALE (p. 73). *Port*, a landing-place; *ing*, a meadow; *scale*, a basin. The place answers this description.

PULL WYKE (p. 43). A *bay* in the *pool* or lake.

SADDLEBACK (p. 85). Explanatory of the outline of the hill; its old name was Blencathra.

SANDWYKE (p. 101). A sandy inlet or bay.

SCANDALE FELL AND BECK (p. 43). *Skans* a fort or rampart,

SCAWFELL (p. 65). (Sax. '*scaew*,' conspicuous) a conspicuous hill—or one that peers above its fellows, as Scawfell does. Or it may be *Scar-fell*.

SCARF-GAP (p. 67). (*Scœf*, Sax. 'smooth') a smooth opening or valley?

SEATHWAITE (p. 13) *Seath* (Sax.) a well or pond, *thwaite*, an inclosure.

SKELWITH (p. 40), *Scale-wath*, a ford in the hollow.

SKIDDAW (p. 84). *Scœd* (Sax.) *sheath*, or *screen;* how (hill). The hill that screens or protects.

STAKE (p. 62). *Stœger* (Sax.) a stair, or road over the hill.

STRIDING EDGE (p. 102). sometimes spelled Strachan Edge. *Strachan* (Sax.) when applied to steps or walking, will be synonymous with 'striding.' *Strid*, a step across. Instance 'Strid,' near Bolton Abbey.

ST. SUNDAY'S CRAG (p. 103). Holy Sunday's Crag—a place where some religious rite has been observed.

STY HEAD (p 63). Stigi, *way*—the head of the way.

THRELKELD (p. 98). *Keld* is a spring of water or well; and *threl* may be corrupted from *Thor's hill*—'Thors hill keld'—and Thirlmere, 'Thor's hill mere.'

TILBERTHWAITE (p. 12). *Till* (Eng.); *bera*, (Sax.) barley; *thwaite*, enclosure. Synonymous with Legberthwaite.

WALLABARROW CRAG (p. 14) *Gwal-beorg*, a natural rampart.

WANSFELL (p. 46). *Wang* (Sax.) a plain field or land. *Wang's-fell*, an exposed hill.

WATENDLATH (p. 75). *Wadan* (Sax.) a ford. *Lathe*, or *lethe*, a district of a country—a 'hundred.'

WHINLATTER (p. 89). *Gwynt-hlaw-tor:* Windy-brow-hill.

WINDERMERE (p. 26). *Gwyn-dwr-mere:* Bright-water-lake.

WRYNOSE (p. 63). The nose of the (*rhiu*) hill.

ULLSWATER (p. 97). BURN derives this name from *Ulf*, *L'ulf*, *Lyulph*, a personal name.

INTRODUCTION.

MR. WEST, in his well-known Guide to the Lakes,* recommends, as the best season for visiting this country, the interval from the beginning of June to the end of August; and the two latter months being a time of vacation and leisure, it is almost exclusively in these that strangers resort hither. But that season is by no means the best: the colouring of the mountains and woods, unless where they are diversified by rocks, is of too unvaried a green; and, as a large portion of the vallies is allotted to hay-grass, some want of variety is found there also. The meadows, however, are sufficiently enlivened after hay-making begins, which is much later than in the southern part of the island. A stronger objection is rainy weather, setting in sometimes at this period with a vigour, and continuing with a perseverance, that may remind the disappointed and dejected traveller of those deluges of rain which fall among the Abyssinian mountains, for the annual supply of the Nile. The months of September and October (particularly October) are generally attended with much finer weather; and the scenery is then, beyond comparison, more diversified, more splendid, and beautiful; but, on the other hand, short days prevent long excursions, and sharp and chill gales are unfavourable to parties of pleasure out of doors. Nevertheless,

* This Guide is now obsolete.

to the sincere admirer of nature, who is in good health and spirits, and at liberty to make a choice, the six weeks following the 1st of September may be recommended in preference to July and August; for there is no inconvenience arising from the season, which, to such a person, would not be amply compensated by the *autumnal* appearance of any of the more retired vallies, into which discordant plantations and unsuitable buildings have not yet found entrance. In such spots, at this season, there is an admirable compass and proportion of natural harmony in colour, through the whole scale of objects; in the tender green of the aftergrass upon the meadows, interspersed with islands of grey or mossy rock, crowned with shrubs or trees; in the irregular inclosures of standing corn, or stubble fields, in like manner broken; in the mountain sides, glowing with fern of divers colours; in the calm blue lakes and river-pools; and in the foliage of the trees, through all the tints of autumn,—from the pale and brilliant yellow of the birch and ash, to the deep greens of the unfaded oak and alder, and of the ivy upon the rocks, upon the trees, and the cottages. Yet, as most travellers are either stinted, or stint themselves, for time, the space between the middle or last week in May, and the middle or last week in June, may be pointed out as affording the best combination of long days, fine weather, and variety of impressions. Few of the native trees are then in full leaf; but, for whatever may be wanting in depth of shade, more than an equivalent will be found in the diversity of foliage, in the blossoms of the fruit-and-berry-bearing trees which abound in the woods, and in the golden flowers of the broom and other shrubs, with which many of the copses are interveined. In those woods, also, and on these mountain-sides which have a northern aspect, and in the deep dells, many of the spring-flowers still linger; while the open and sunny places are stocked with the flowers of the approaching summer. And, besides, is not an exquisite pleasure still untasted by him who has not heard the choir of linnets and thrushes chaunting their love-songs in the copses, woods, and hedge-rows of a mountainous country; safe from the birds of prey, which build in the inaccessible crags, and are at all hours seen or heard wheeling about in the air? The number of these formidable creatures is probably the chief cause, why, in the *narrow* vallies, there are no skylarks; as the destroyer would be enabled to dart upon them from the surrounding crags, before they could descend to their ground-nests for protection. It is not often that the nightingale resorts to these vales; but almost all the other tribes of our English warblers are numerous; and their notes, when listened to by the side of broad still-waters, or when heard in unison with the murmuring of mountain-brooks, have the compass of their power enlarged accordingly. There is also an imaginative influence in the voice of the cuckoo, when that voice has taken possession of a deep mountain valley, very different from any thing which can be excited by the same sound in a flat country. Nor must a circumstance be omitted, which here renders the close of spring especially interesting; I mean the practice of bringing down the ewes from the mountains to yean in the vallies and enclosed grounds. The herbage being thus cropped as it springs, *that* first tender emerald green of the season, which would otherwise have lasted little more than a fortnight, is prolonged in the pastures and meadows for many weeks; while they are farther enlivened by the multitude of lambs bleating and skipping about. These sportive creatures, as they gather strength, are turned out upon the open mountains and, with their slender limbs, their snow-white colour, and their wild and light motions, beautifully accord or contrast with the rocks and lawns, upon which they must now begin to seek their food. And last, but not least, at this time the traveller will be sure of room and comfortable accommodation, even in the smaller inns. I am aware that few of those who may be inclined to profit by this recommendation will be able to do so, as the time and manner of an excursion of this kind are mostly regulated by circumstances which prevent an entire freedom of choice. It will therefore be more pleasant to observe, that, though the months of July and August are liable to many objections, yet it often happens that the weather, at this time, is not more wet and stormy than they—who are really capable of enjoying the sublime forms of nature in their utmost sublimity—would desire. For no traveller, provided he be in good health, and with any command of time, would have a just privilege to visit such scenes, if he could grudge the price of a little confinement among them, or interruption in his journey, for the sight or sound of a storm coming on or clearing away. Insensible must he be who would not congratulate himself upon the bold bursts of sunshine, the descending vapours, wandering lights and shadows, and the invigorated torrents and waterfalls, with which broken weather, in a mountainous region, is accompanied. At such a time there is no cause to complain, either of the monotony of midsummer colouring, or the glaring atmosphere of long, cloudless, and hot days.

Thus far concerning the respective advantages and disadvantages of the different seasons for visiting this country. As to the order in which objects are best seen—a lake being composed of water flowing from higher grounds, and expanding itself till its receptacle is filled to the brim,—it follows, that it will appear to most advantage when approached from its outlet, especially if the lake be in a mountainous country; for, by this way of approach, the traveller faces the grander features of the scene, and is gradually conducted into its most sublime recesses. Now, every one knows, that from amenity and beauty the transition to sublimity is easy and favourable; but the reverse is not so; for, after the faculties have been elevated, they are indisposed to humbler excitement.*

It is not likely that a mountain will be ascended without disappointment, if a wide range of prospect be the object, unless either the summit be reached before sunrise, or the visitant remain there until sun-set, and afterwards. The precipitous sides of the mountain, and the neighbouring summits, may be seen with effect under any atmosphere which allows them to be seen at all; but *he* is the most fortunate adventurer, who chances to be involved in vapours which open and let in an extent of country partially, or, dispersing suddenly, reveal the whole region from centre to circumference.

A stranger to a mountainous country may not be aware that his walk in the early morning ought to be taken on the eastern side of the vale, otherwise he will lose the morning light, first touching the tops and thence creeping down the sides of the opposite hills, as the sun ascends, or he may go to some central

* The only instance to which the foregoing observations do not apply, are Derwent Water and Lowes Water. Derwent is distinguished from all the other Lakes by being *surrounded* with sublimity: the fantastic mountains of Borrowdale to the south, the solitary majesty of Skiddaw to the north, the bold steeps of Wallow Crag and Lodore to the east, and to the west the clustering mountains of Newlands. Lowes Water is tame at the head, but towards its outlet has a magnificent assemblage of mountains. Yet as far as respects the formation of such receptacles, the general observation holds good: neither Derwent nor Lowes Water derive any supplies from the streams of those mountains that dignify the landscape towards its outlets.

eminence, commanding both the shadows from the eastern, and the lights upon the western, mountains. But, if the horizon line in the east be low, the western side may be taken for the sake of the reflections, upon the water, of light from the rising sun. In the evening, for like reasons, the contrary course should be taken.

After all, it is upon the *mind* which a traveller brings along with him that his acquisitions, whether of pleasure or profit, must principally depend.—May I be allowed a few words on this subject?

Nothing is more injurious to genuine feeling than the practice of hastily and ungraciously depreciating the face of one country by comparing it with that of another. True it is, "Qui *bene* distinguit bene *docet*;" yet fastidiousness is a wretched travelling companion; and the best guide to which, in matters of taste we can entrust ourselves, is a disposition to be pleased. For example, if a traveller be among the Alps, let him surrender up his mind to the fury of the gigantic torrents, and take delight in the contemplation of their almost irresistible violence, without complaining of the monotony of their foaming course, or being disgusted with the muddiness of the water—apparent even where it is violently agitated. In Cumberland and Westmorland, let not the comparative weakness of the streams prevent him from sympathysing with such impetuosity as they possess; and, making the most of the present objects, let him, as he justly may do, observe with admiration the unrivalled brilliancy of the water, and that variety of motion, mood, and character, that arises out of the want of those resources by which the power of the streams in the Alps is supported.—Again, with respect to the mountains; though these are comparatively of diminutive size, though there is little of perpetual snow, and no voice of summer avalanches is heard among them; and though traces left by the ravage of the elements are here comparatively rare and unimpressive, yet out of this very deficiency proceeds a sense of stability and permanence that is, to many minds, more grateful—

> " While the coarse rushes to the sweeping breeze
> Sigh forth their ancient melodies."

Among the Alps are few places which do not preclude this feeling of tranquil sublimity. Havoc, and ruin, and desolation, and encroachment, are everywhere more or less obtruded; and it is difficult, notwithstanding the naked loftiness of the *pikes*, and the snow-capped summits of the *mounts*, to escape from the depressing sensation, that the whole are in a rapid process of dissloution; and, were it not that the destructive agency must abate as the heights diminish, would, in time to come, be levelled with the plains. Nevertheless, I would relish to the utmost the demonstrations of every species of power at work to effect such changes.

From these general views let us descend a moment to detail. A stranger to mountain imagery naturally, on his first arrival, looks out for sublimity in every object that admits of it; and is almost always disappointed. For this disappointment there exists, I believe, no general preventive; nor is it desirable that there should. But with regard to one class of objects, there is a point in which injurious expectations may be easily corrected. It is generally supposed that waterfalls are scarcely worth being looked at except after much rain, and that, the more swoln the stream the more fortunate the spectator; but this, however, is true only of large cataracts with sublime accompaniments: and not even of these without some drawbacks. In other instances, what becomes, at such a time, of that sense of refreshing coolness which can only be felt in dry and sunny weather, when the rocks, herbs, and flowers glisten with moisture diffused by the breath of the precipitous water? But, considering these things as objections of sight only, it may be observed that the principal charm of the smaller waterfalls or cascades consists in certain proportions of form and affinities of colour, among the component parts of the scene; and in the contrast maintained between the falling water and that which is apparently at rest, or rather settling gradually into quiet in the pool below. The beauty of such a scene, where there is naturally so much agitation, is also heightened, in a peculiar manner, by the *glimmering*, and, towards the verge of the pool, by the *steady* reflection of the surrounding images. Now, all those delicate distinctions are destroyed by heavy floods, and the whole stream rushes along in foam and tumultuous confusion. A happy proportion of component parts is indeed noticeable among the landscapes of the North of England; and, this characteristic, essential to a perfect picture, they surpass the scenes of Scotland, and, in a still greater degree, those of Switzerland.

DIRECTIONS AND INFORMATION

FOR

THE TOURIST.

THE DISTRICT OF THE LAKES is now so conveniently approached from all quarters by railway, that the Routes formerly laid down are no longer considered applicable for the generality of Tourists.

Commencing at Preston, there are two approaches to the Lakes from the south, the only direct one being that by way of KENDAL to WINDERMERE, a small town rapidly rising into importance at the terminus of the Kendal and Windermere Railway, from whence the Lake Visitor may, with the greatest convenience, commence his tour. The other is through FURNESS, which is gained by diverging to the west at PRESTON for FLEETWOOD, or at LANCASTER for POULTON, and crossing the estuary by steamboats from either place, to Furness Abbey and Ulverston; thence to BOWNESS, on the banks of Windermere.

Parties from Yorkshire would find Lancaster a more convenient point of divergence than Preston, and the time occupied in crossing to the Furness coast would be about the same.

Travellers from the North would do well to go from CARLISLE to Maryport by railway, and proceed by Cockermouth and along the banks of Bassenthwaite to KESWICK; or they may proceed from Carlisle to Penrith, and thence cross the country to Keswick, and begin with that vale, rather than with Ullswater, taking Patterdale and Ullswater on their way to Ambleside and the South.

We purpose, first, to point out the approach to the Lake District by the FURNESS ROUTE, as far as BOWNESS and the village of WINDERMERE, and afterwards conduct the Tourist to the same place by way of KENDAL. We shall thence direct him to AMBLESIDE and KESWICK, as being the most important Stations from whence to make Excursions.

FURNESS ROUTE.

The distance from Preston to Fleetwood, by rail, is accomplished in one hour, and another hour will land the tourist at the harbour of Piel, on the opposite coast, whence there is a train, on the arrival of every steamer, to FURNESS ABBEY, about four miles distant, and thence by Dalton to Ulverston. PIEL CASTLE, on the Isle of Walney, will be noticed at a short distance from the pier, on landing. It was erected by one of the Abbots of Furness in the the time of Edward III.

FURNESS ABBEY

Possesses peculiar attractions to the antiquarian and the pleasure-seeker; and, being now so easily approached, is a place of great resort. It is the property of the Earl of Burlington, who has, since the introduction of the railway, which passes through a part of the ruins, converted the Abbot's house into a commodious hotel, and laid out the area adjoining as a pleasure-ground, in a style according well with the monastic character of the place.

The Monastery, according to the authority of John Stell, a Monk who belonged to the House, was first planted at Tulket, in Amounderness, in the year 1124; three years after which, viz. on the 1st of July, 1127, it was translated, and founded by Stephen, Earl of Bologna and Morton (afterwards King of England), in the vale of Bekansgill,* in the Peninsula of Furness.

Furness is an abbreviation of *Frudernesse* (as the name appeared in Doomesday Book), or *Futhernesse*, as it seems to have been more frequently written. *Futher* is conjectured by Dr. Whitaker to be a personal name, probably that of the first Saxon planter or proprietor of the district: *Nesse* is a promontory; than which hardly any appellation could be more appropriate, as descriptive of the southern extremity of the territory where the Abbey stands.

The Monks of Furness originally belonged to the Savignian order; an order which, of all others, complied most scrupulously

* Bekansgill, from Lethel Bekan, the *Solanum Lethale*, or Deadly Night Shade, which once abounded in the district.

with the rules of the great parent of monachal institutions, St. Benedict. About 1148, in the Pontificate of Eugenius III.; the whole order of Savignian Monks matriculated into the Cistercian or Bernardine, in honour of St. Bernard, a man of great sanctity and learning, who reformed and remodelled the Benedictine rules. In the time of Bajocis, their *fifth* Abbot, the Monks of Furness (after some hesitation and opposition) consented to become Cistercians, the rules of which order they religiously observed till the general Dissolution of Monasteries.

Rising from its titular Saint, Bernard, and twelve monks, who filiated from Citeaux,* the Cistercian order, in an incredibly short time, became of great repute and corresponding extent. So rapid was its progress that before the death Saint Bernard, he had founded 160 Monasteries; and in the space of fifty years from its first establishment as an order, it had acquired 800 Abbeys! All the Houses belonging to this Order were dedicated to the *Virgin Mary*.

In England and Wales there were eighty-five Houses of the Cistercian order; of which number two only were situated in the County of Lancaster, viz. Furness and Whalley. Until the time of Pope Sextus IV. the rules and observances, both as to fasting and religious devotions, were uncommonly rigorous; but this Pontiff published a decree to mitigate the austerities of their spiritual exercises, and to preserve uniformity in table and dress. From this time they were allowed to eat flesh three times in a week, for which purpose a particular dining room, distinct from the usual Refectory, was fitted up in every Monastery.

Their dress was a white† Cassock, with a Caul and Scapulary of the same. For the Choir dress they wore a white or grey Cassock, with Caul and Scapulary of the same, and a girdle of black wool; over that a Mozet, or Hood, and a Rocket, the front part of which descended to the girdle, where it ended in a round, and the back part reached down to the middle of the leg behind. Whenever they appeared abroad, they wore a Caul and a full black Hood. This is only a general description of their dress; for every House had something particular to itself.

* Hence the name of the order, *Cistercian*.

† The dress of the Savignians was *grey*, from which they were usually called Grey Monks.

With respect to the power, privileges, benefactions and possessions of Furness Abbey, it would take almost an entire volume fully to narrate and illustrate the whole.

The Lordship of Furness comprehends all that tract of land, with the islands included, commencing in the north at the Shire Stones, on Wrynose Hills, and descending by Elterwater into Windermere, and by the outlet of that lake at Newby Bridge, over Leven Sands into the sea. Extending along the sea, it includes the isle of Foulney, the pile of Fouldrey, and the Isle of Walney. Beyond which, turning to the north-east, it ascends, first by the estuary of Duddon, and then by the river itself,—which, by the names of Duddon and, higher up, of Cockley Beck, traces an ascending line to Shire Stones again, where the boundary commenced.

The power of the Abbot, throughout the whole of this territory, in affairs both ecclesiastical and civil, was confessedly absolute. Within these limits he exacted the same oath of fealty which was paid to the King. The veneration which the sanctity and dignity of his office inspired, and the circumstance of his territory being bounded on one hand by seas almost impassable, and on the other by mountains almost insurmountable, conspired to give to Furness the character and importance of a separate and independent kingdom. Even the military establishment of the district depended upon the Abbot; and every Mesne Lord obeyed his summons in raising his quota of armed men for guarding the coasts or for the border service. He had the patronage of all the Churches, except one. He had also, by prescription, the appointment of Coroner and Chief Constable, and all Officers incident to the Courts Baron. He, and all his men, were free from all county amerciaments, and suits of counties and wapentakes. He had a free market and fair in Dalton; with a court of criminal jurisdiction. He issued summonses and attachments by his own bailiffs. He had the return of all writs; and the Sheriff, with his officers, were prohibited from entering his territories under any pretext of office whatever. His lands and tenants were exempt from all legal exactions of talliage, toll, passage, pontage, and vectigal; and no man was to presume to disturb or molest the Abbot, or any of his tenants, on pain of forfeiting ten pounds to the King! In addition to all which he was immediate owner and occupant of almost half the low country. And for protections, privileges, and immunities, there were few Monasteries indeed that could boast so much. Pope Eugenius III. and Pope Innocent III. both conferred special favours on the Furness Monks; and the princely foundation of STEPHEN was confirmed and secured to them by the Charters of twelve succeeding Monarchs of England. Immense wealth was, besides, conferred on them by propitiatory offerings of the neighbouring families of opulence, who consecrated their substance with their bodies to the sacred retirement of the Abbey.

With these means and appliances, the Monks exercised absolute dominion over the whole peninsula of Furness during four centuries, from the foundation of the Abbey till the general dissolution of Monasteries in the time of Henry VIII., when all power and authority, wealth and honours, were surrendered up to the King. The last Abbot was humbled to accept, as a pension, during the remainder of his life, the profits of the Rectory of Dalton, which were then valued at £33 6s. 8d. per annum.

Such is a brief and bare outline of the history of this once great and magnificent Abbey. The situation of the Monastery indicates the peculiar good taste of the architects. Secluded in a deep glen, which nevertheless opens out below into an expanse of fertile meadows, irrigated by a murmuring brook, and screened by a forest of stately timber, the contemplative Monks could here, unawed and unseen, perform their holy rites, and pour out their souls in prayer!

> "Such is the dwelling, grey and old, which in some world-worn mood,
> The youthful poet dreamed would suit his future solitude;
> If the old abbey be his search, he might seek far and near
> Ere he could find a gothic Cell more lonely than was here.
> Long years have darkened into time since Vespers here were rung,
> And here has been no other dirge than what the winds have sung
> And now the drooping ivy wreaths in ancient clusters fall,
> And moss o'er each device hath grown upon the sculptured wall."

We find nothing to add to Mr. West's description of the edifice in the "Antiquities of Furness," published in 1805. The ruins since that time have undergone very little alteration:—

The magnitude of the Abbey may be known from the dimensions of the ruins; and enough is standing to show that in the style of architecture prevailed the same simplicity of taste which is found in most houses belonging to the Cistercian monks, which were erected about the same time with Furness Abbey. The round and pointed arches occur in the doors and windows. The fine clustered Gothic and the heavy plain Saxon pillars stand contrasted. The walls shew excellent masonry, are in many places counterarched, and the ruins discover a strong cement.

The east window of the church has been noble; some of the painted glass that once adorned it is preserved in a window in Bowness Church. The window consists of seven compartments, or partitions. In the third, fourth, and fifth, are depicted, in full proportion, the Crucifixion, with the Virgin Mary on the right, and the beloved disciple on the left side of the cross: angels are expressed receiving the sacred blood from the five precious wounds: below the cross is a group of Monks in their proper habits, with the abbot in a vestment; their names are written on labels issuing from their mouths; the abbot's name is defaced, which would have given a date to the whole. In the second partition are the figures of St. George and the dragon. In the sixth is represented St. Catharine, with the emblems of martyrdom, the sword and wheel. In the seventh are two figures of mitred abbots, and underneath them two monks dressed in vestments. In the middle compartment, above, are finely painted, quarterly, the arms of France and England, bound with the garter and its motto, probably done in the reign of Edward III. The rest of the window is filled up by pieces of tracery, with some figures in coats armorial, and the arms of several benefactors, amongst whom are Lancaster, Urswick, Harrington, Fleming, Millum, &c.

On the outside of the window at the Abbey, under an arched festoon, is the head of Stephen the founder: opposite to it, that of Maude his queen, both crowned, and well executed. In the south wall, and east end of the church, are four seats adorned with Gothic ornaments. In these the officiating priest, with his attendants, sat at intervals, during the solemn service of high mass. In the middle space, where the first barons of Kendal are interred, lies a procumbent figure of a man in armour, cross-legged.

The chapter-house is the only building belonging to the Abbey which is marked with any elegance of Gothic sculpture; it has been a noble room of sixty feet by forty-five. The vaulted roof, formed of twelve ribbed arches, was supported by six pillars in two rows, at fourteen feet distance from each other. Now, supposing each of the pillars to be eighteen inches in diameter, the room would be divided into three alleys, or passages, each fourteen feet wide. On entrance, the middle one only could be seen, lighted by a pair of tall pointed windows at the upper end of the room; the company in the side passage would be concealed by the pillars, and the vaulted roof, that groined from these pillars, would have a truly Gothic disproportioned appearance of sixty feet by fourteen. The northern side alley was lighted by a pair of similar side lights, and a pair at the upper end: the southern side alley was lighted by four small pointed side windows, besides a pair at the higher end, at present entire, and which illustrate what is here said. Thus, whilst the upper end of the room had a profusion of light, the lower end would be in the shade. The noble roof of this singular edifice did but lately fall in: the entrance or porch is still standing, a fine circular arch, beautified with a deep cornice, and a portico on each side. The only entire roof of any apartment now remaining, is that of a building without the enclosure wall, which is supposed to have been a private chapel to the Guest-Hall. It is a single-ribbed arch that groins from the wall.

The tower has been supported by four magnificent arches, of which only one remains entire. They rested upon four tall pillars, whereof three are finely clustered, but the fourth is of a plain unmeaning construction.

The west end of the church seems to have been an additional part, intended for a belfry, to ease the main tower; but that is as plain as the rest: had the monks even intended it, the stone would not admit of such work as has been executed at Fountains and Rievaulx Abbies. The east end of the church contained five altars, besides the high altar, as appears by the chapels; and probably there was a private altar in the sacristy. In magnitude, this Abbey was the second in England belonging to the Cistercian monks, and next in opulence after Fountains Abbey, in Yorkshire. The church and cloisters were encompassed with a wall, which commenced at the east side of the great northern door, and formed the strait enclosure; and a space of ground, to the amount of sixty-five acres, was surrounded with a stone wall, which enclosed the mills, kilns, ovens, and fish-ponds belonging to the Abbey, the ruins of which are still visible. This last was the great enclosure, now called the Deer Park, in which such terraces might be formed as would equal, if not surpass, any in England.

EXPLANATION OF THE GROUND-PLAN OF FURNESS ABBEY.

A, B, C, Q, T, V, N, represent the parts of the church.

A, the east end of the church, where the higher altar stood. Behind that was the circumambulatory.

In the south wall was placed the piscina, or cistern, at which the priest washed his hands before service; there is also a small niche, and over it hung the manutergium, on each side of the cistern, for receiving the purifactories. Below these are four stalls, or seats, in the wall, richly ornamented in the Gothic style, in which the officiating priest, with his assistants, sat at intervals, in time of celebrating high mass.

Q, the choir.—CC, chapels.—V. vestry.

TT, the transept. At the north end of the transept below T, is the grea door into the church; and at the south end is a door-case leading to the dormitory, through which the monks came into the church at midnight to sing matins, or morning prayers. On the west side of the door at the north end of the transept, there is a spiral stair--case, which, after rising in a perpendicular direction for a considerable height, has branched out into a passage in the western wall, and led to another flight of spiral stairs, on the top of one of the clustered columns, which supported the central spire over the intersection of the nave and transept. These different flights of steps have formed the communication between the ground floor of the church and the higher parts of the spire.

N, the nave of the church. Above N, is the southern aisle: and below N, is the northern aisle. In the south wall adjoining the transept, is a door-way opening into a quadrangular court. There has probably been also a door-way in the north-wall, near the west end of the nave.

B, the belfry, or tower, at the west end of the church. In the wall on the south side of the ruins of this tower, close to the west window, there is a part of the spiral stairs which led to the top of the tower.

CH, CL, H, K, L, M, NO, O, P, PL, QC, R, S, U, represent the chapter-house, the cloisters, and part of the Abbey adjoining.

CH, the chapter-house, over which were the library and scriptorium. The roof is represented as it lately stood. The porch has been ornamented with a deep ox-eye cornice, and pilastres of marble. The pilasters are demolished, but the roof is entire. On each side of this porch there is a portico in the wall, with a similar cornice.

R, the dining-room, or refectory. There has been a passage leading from it to K, the kitchen and offices, over which were lodging-rooms for the secular servants.

L, the locutorium, the calefactory, and conversation room.

H, halls and rooms.

S, a building on the outside of the strait enclosure, supposed, by West, to have been the school-house, but now generally admitted to have been a private chapel to the Guest-Hall. There is a stone seat all round, and in the south wall is the stone pillar upon which was erected the pulpit of the preacher. The roof of this building is entire, and also that of a passage adjoining. Over these have been apartments.

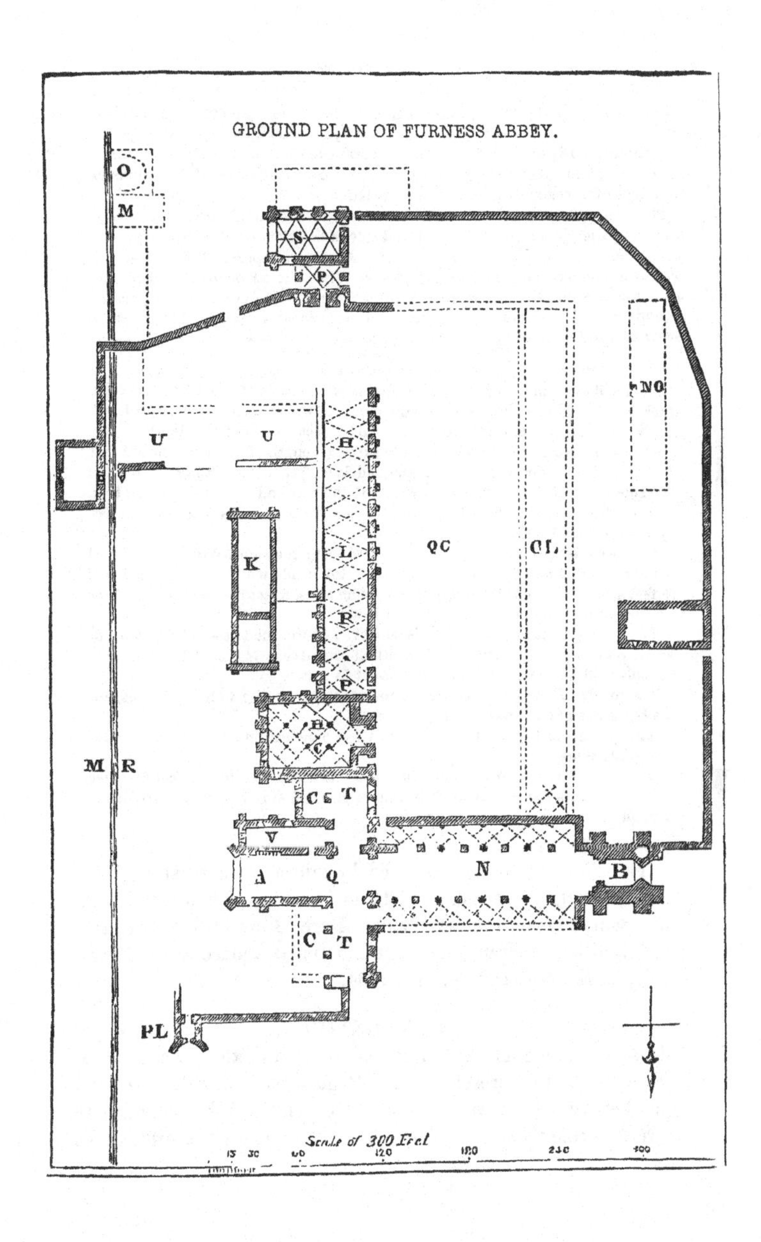
GROUND PLAN OF FURNESS ABBEY.
O
M
S
P
U
U
H
NO
QC
CL
K
L
R
P
H
C
M R
C & T
V
A
Q
N
B
C
T
PL
Scale of 300 Feet

PP, passages.—CL, the opposite wing of the cloisters razed to the ground.—QC, the area of the quadrangular court.—PL, a porter's lodge and gateway.—M, the mill.—MR, the mill-race.—O, the great oven.—NO, the ruins of a building of uncertain extent, supposed to have been the novitiate.—UU, the ruins of buildings of uncertain extent and appropriation.

The rivulet from the north, which constantly runs through the valley, is conducted by the east end of the church and side of the cloisters in a subterraneous passage or tunnel, which is arched over. Another temporary brook from the west, has been conducted by NO, and under S, in a similar manner. There has also been a subterraneous passage, leading from the race of the rivulet, under K, and forwards in an unknown direction. It has probably been conducted under some part of the church, and has served for a drain or sewer.

DIMENSIONS OF THE CHURCH, THE CHAPTER-HOUSE, AND CLOISTERS.

The inside length of the church, from east to west, is 275 feet 8 inches: the thickness of the east end wall, and the depth of the east end buttress, 8 feet 7 inches: the thickness of the west end wall, 9 feet 7 inches: the depth of the west end buttress, 10 feet 8 inches: the extreme length of the church, 304 feet 6 inches. The inside width of the east end is 28 feet, and the thickness of the two side walls, 10 feet. The total width of the east end is, therefore, 38 feet. The height of the arch above Q, from the floor to the underside of the centre-stone, is 52 feet 6 inches.

The inside length of the TRANSEPT is 130 feet: the south-wall is 6 feet, and the north wall 3 feet 6 inches in thickness: the inside width of the transept is 28 feet 4 inches: the thickness of the two side walls, 8 feet 8 inches. The whole breadth of the transept is, therefore, 37 feet.

The inside width of the nave is 66 feet; and the thickness of the two side walls, 8 feet: therefore the whole width of the nave is 74 feet. The height of the side walls of the church has been about 54 feet.

The inside of the CHAPTER-HOUSE measures 60 feet by 45 feet 6 inches, and the thickness of each wall, 3 feet 6 inches.

The inside width of the CLOISTERS is 31 feet 6 inches, and the thickness of the two walls, 8 feet.

The area of the quadrangular court is 338 feet 6 inches by 102 feet 6 inches. On solemn days the monks used to walk in procession round this court, under a shade.

The Tourist must now proceed by railway to Ulverston, passing DALTON, the ancient Capital of Furness, with a population of about 800, on the left hand. Here GEORGE ROMNEY, the distinguished portrait painter, was born, at a place called Beckside, on the 5th of December, 1734.

ULVERSTON

Is a flourishing market-town and port, and the emporium of Furness at the present day. Population, about 5,000, and market-day Thursday. Considerable quantities of iron and slate are exported from this place. There are many beautiful walks in the neighbourhood, and particularly in the grounds of CONISHEAD PRIORY. Inns—*Sun* and *Bradyll's Arms*.

From the HILL OF HAUD there is an extensive prospect, and on its summit a magnificent column has recently been erected to the memory of Sir John Barrow, one of the Secretaries to the Admiralty, who was born at DRAGLEY BECK, close by. It is a commanding object for many miles round.

SWART-MOOR HALL may also be mentioned, as once the residence of Judge Fell, whose widow married George Fox, a leader amongst the Quakers at that period. It is now a dilapidated farm-house, and possesses no interest except what attaches to it from the above circumstance. The Friends have a meeting-house at Swart-Moor, which was built by Fox, and was the first place of religious worship erected for the use of that community.

From Ulverston the Lakes would be advantageously approached by Coniston; thence to Hawkshead, and by the Ferry over Windermere, to Bowness. Or, the Tourist may, by leaving out Coniston, proceed direct to Bowness, by way of Newby Bridge, at the foot of Windermere, eight miles from Ulverston, where there is a capital inn, and from whence steam-boats ply regularly during the summer season to all the Stations on the Lake, at very moderate fares.

Should Coniston be adopted, the road is along a narrow vale, beautifully diversified by hanging inclosures and scattered farms, half way up the sides of the mountains, whose heads are covered with heath and brown vegetation. About three miles from Ulverston observe a farm-house on the left, and a group of houses before you on the right. Stop at the gate on the brow of the hill, and have a distant view of the lake. The whole range of Coniston fells is now in sight. Advancing, on the left see Lowick Hall, once the seat of a family of that name. Cross the river Crake at Lowick, and keep on the eastern side of the lake of Coniston till you reach the inn at its head. The distance is sixteen miles.

Excursions from Coniston Water Head Inn.

This Inn has lately been rebuilt, a little to the south of the old site, in a style of great magnificence, and every accommodation is afforded to travellers visiting this interesting part of the District. From it several delightful excursions might be made, and the tourist would act wisely in taking up his abode here for a few days.*

A leisurely traveller might have much pleasure in looking into Yewdale and Tilberthwaite, returning from the head of Yewdale by a mountain track which has the farm of Tarn Hows a little on the right. By this road is seen much the best view of Coniston Lake from the north.

An enterprising tourist might go to the Vale of Duddon, over Walna Scar, down to Seathwaite, Newfield, and to the rocks where the river issues from a narrow pass into the broad vale. Horses may be taken over this mountain track, which is, however, in places very steep and difficult. The distance is six miles.

The stream is very interesting for the space of a mile above this point, and below, by Ulpha Kirk, till it enters the Sands, where it is overlooked by the solitary mountain Black Comb, the summit of which, as that experienced surveyor, Colonel Mudge, declared, commands a more extensive view than any point in Britain. Ireland he saw more than once, but not when the sun was above the horizon.

"Close by the Sea, lone sentinel,
  Black-Comb his forward station keeps:
He breaks the sea's tumultuous swell,—
  And ponders o'er the level deeps.
He listens to the bugle horn,
  Where Eskdale's lovely valley bends,
Eyes Walney's early fields of corn
  Sea-birds to Holker's woods he sends.
Beneath his feet the sunk ship rests,
In Duddon Sands, its masts all bare."
* * * * * * * * *

The Minstrels of Windermere, by Chas. Farish, B. D.

* A full and accurate description of this and the neighbouring vales has recently been published in a handsome little volume, entitled "THE OLD MAN, or Ravings and Ramblings round Coniston," which is on sale, we believe, at the Post-office in the village.

The carriage-road to Seathwaite is by either of the two following routes, and affords many pleasing and extensive prospects:—

1	Coniston Church	1	1	Duddon Bridge	11½
2½	Torver	3½	3½	Ulpha Kirk-house	15
7	Broughton	10½	2	Newfield, near Seathwaite Chapel	17

OR,

3½	Torver	3½	2	Broughton Mills	8½
3	Three miles beyond Torver take the road to the right	6½	4	Newfield	12½

The following description of the scenery in this Excursion is extracted from Mr. Wordsworth's Notes to the river Duddon:

"This recess (the Vale of Seathwaite), towards the close of September, when the after-grass of the meadows is still of a fresh green, with the leaves of many of the trees faded, but perhaps not fallen, is truly enchanting. At a point elevated enough to show the various objects in the valley, and not so high as to diminish their importance, the stranger will instinctively halt. On the foreground, a little below the most favourable station, a rude foot-bridge is thrown over the bed of the noisy brook foaming by the way side. Russet and craggy hills, of bold and varied outline, surround the level valley, which is besprinkled with grey rocks plumed with birch trees. A few homesteads are interspersed, in some places peeping out from among the rocks like hermitages, whose sites have been chosen for the benefit of sunshine as well as shelter; in other instances, the dwelling-house, barn, and byre, compose together a cruciform structure, which, with its embowering trees, and the ivy clothing part of the walls and roof like a fleece, call to mind the remains of an ancient abbey. Time, in most cases, and nature everywhere, have given a sanctity to the humble works of man, that are scattered over this peaceful retirement. Hence a harmony of tone and colour, a consummation and perfection of beauty, which would have been marred had aim or purpose interfered with the course of convenience, utility, or necessity. This unvitiated region stands in no need of the veil of twilight to soften or disguise its features. As it glistens in the morning sunshine, it would fill the spectator's heart with gladsomeness. Looking from our chosen station, he would feel an impatience to rove among its pathways, to be greeted by the milkmaid, to wander from house to house, exchanging 'good-morrows' as he passed the open doors; but, at evening, when the sun is set, and a pearly light gleams ftom the western quarter of the sky, with an unanswering light from the smooth surface of the meadows; when the trees are dusky; but each kind still distinguishable; when the cool air has condensed the blue smook rising from the cottage chimneys; when the dark mossy stones seem to sleep in the bed of the foaming brook; *then*, he would be unwilling to move forward, not less from a reluctance to relinquish what he beholds, than from an apprehension of disturbing, by his approach, the quietness beneath him. Issuing from the plain of this valley, the brook descends in a rapid torrent, passing by the church-yard of Seathwaite. From the point where the Seathwaite brook joins the Duddon, is a view upwards, into the pass through which the river makes its way into the plain of Donnerdale. The perpendicular rock on the right bears the ancient name of THE PEN; the one opposite is called WALLABARROW CRAG, a name that occurs in other places, to designate rocks of the same character. The *chaotic* aspect of the scene is well marked by the expression of a stranger who strolled out while dinner was preparing, and, at his return, being asked by his host, 'What way he had been wandering?' replied, 'As far as it is *finished!*'"*

* Seathwaite is remarkable as the place in which "Wonderful Robert Walker, dwelt for the greatest part of a century. A very full and interesting account of this extraordinary man is given by Mr. Wordsworth in his Notes to "The Duddon," to which work the reader is referred. It may here suffice to say, that he was born in 1709, at Under-Crag, in Seathwaite, and was the youngest of twelve children. Being sickly in youth, he was "bred a scholar," and, after acting for some time as a schoolmaster at Loweswater, in Cumberland, he was ordained, and, about 1735, became curate of Seathwaite, where he remained till his death, sixty-seven years afterwards. The value of his curacy when he entered upon it was £5 per annum, with a cottage. About the same time he married, and his wife brought him, as he says, "to the value of £40 to her fortune." He had a family of twelve children, of whom, however, only eight lived: these he educated respectably, and one of his sons became a clergyman. He was even munificent in his hospitality as a parish priest, and generous to the needy; and although the income of his curacy never exceeded £50 per annum, he "at his decease left behind him no less a sum than £2,000; and such a sense of his various excellencies was prevalent in the country, that the epithet of WONDERFUL is to this day attached to his name." He died on the 25th of June, 1802, in the 93rd year of his age, and 67th of his curacy. His wife died on the 28th of January in the same year, and at the same age.

The Tourist may either return to the inn at Coniston, or he may cross from Ulpha Kirk over Birker Moor to Ambleside, by the following route; —

4 Stanley Gill	4	16 Ambleside, over Hardknot and Wrynose	22
2 Birker Force	6		

After leaving Ulpha Kirk, he should, in proceeding over the moor, take care to turn to the right by a very indifferent road (apparently leading only to a farm-house), before beginning to descend into Eskdale, which will conduct him to STANLEY GILL, at the head of the finest ravine in the country. Three-quarters of a mile higher up the valley, on the same side, appears BIRKER FORCE, dashing over a high, naked, and precipitous rock.* Thence proceed up the Vale of the Esk, by Hardknott and Wrynose, to Ambleside. Near the road, in ascending from Eskdale, are conspicuous remains of a Roman fortress, called by the country people "*Hardknott Castle,*" most impressively situated on the left, half way up the hill. It has escaped the notice of most antiquarians, and is but slightly mentioned by Lysons. There is a DRUIDICAL CIRCLE about half a mile to the left of the road ascending Stoneside from the Vale of Duddon: the country people call it "*Sunken Church.*" The road over Hardknott and Wrynose is scarcely practicable except on foot or on horseback.

The ascent to the top of the OLD MAN Mountain is recommended before leaving Coniston; but the ground being rugged, in places, it should not be undertaken without a guide. The height of the Old Man is 2577 feet, and the view from it is inferior to no mountain view in the country, excepting that from Scawfell or Helvellyn, if indeed it be inferior to the latter. The ascent should be made by following the ancient horse-road over Walna Scar for about a mile, and then turning to the right towards an old slate-quarry, whence you will have to scramble to the summit. LOW WATER lies immediately below the highest point, in a hollow of the mountain, to the east, and GOAT'S

* STANLEY GILL is often erroneously called BIRKER FORCE by the dalesmen, by which confusion of the two names the stranger is apt to be misled. The original name of this fall was, we believe, Dalegarth Force; and was changed to Stanley Gill by the present proprietor, Mr. Stanley, of Ponsonby.

WATER is situated under the precipitous side of Dow Crag on the west. The stream from it flows into Coniston. BLIND TARN (so called, perhaps, from its having no outlet) will be seen further to the south, under a part of Walna Scar. A walk of half a mile from the top towards the north-west will bring the traveller in sight of SEATHWAITE TARN, which sends a tributary to the Duddon. Those who can give a day to the excursion will do well to follow the mountain range to Wetherlam, a lofty ridge that sweeps round to the north of the Old Man, under which lies a fine Tarn called LEVERS WATER, where copper-mining is carried on much to the injury of this magnificent scene. From Wetherlam descend into Tilberthwaite, and so return to Coniston.

The LAKE OF CONISTON is six miles long and three-quarters of a mile in breadth. Its greatest depth is twenty-seven fathoms, and it is famous for its charr (*salmo alpinus*), a species of trout, which inhabits the deep water, and is only taken at particular times of the year. Large quantities are potted and sent to the south. They do not attain a large size, seldom, perhaps, exceeding a pound in weight. Coniston, Windermere, Wastwater, Buttermere, Crummock, and Ullswater, are, it is said, the only lakes which contain them. The charr of Coniston Water stand highest, and those of Ullswater lowest, in repute.

The road from Coniston Water Head to Ambleside direct, is eight miles; but, as has been before said, a circuitous route by Hawkshead, the Ferry, and Bowness, 15 miles, in the following order, is recommended as a much better introduction to Windermere.

HAWKSHEAD

Is a compact little market-town, at the southern end of which, on a good elevation, stands the Parish Church, commanding a pleasant prospect of the Vale and Lake of ESTHWAITE, the latter of which is two miles long and half a mile in breadth. Here is a Free Grammar School, founded in 1585, by Edwin Sandys, Archbishop of York, whose family is yet found in the vicinity. Some years ago this school was filled with pupils not only from the neighbourhood but from the surrounding counties, numbering at one period about 120. The poet Wordsworth, and the late Dr. Wordsworth, his brother, with many others distinguished for classical attainments, were educated here.

There is a pleasant drive round Esthwaite Water by the Grove and Esthwaite Hall, passing ESTHWAITE LODGE (Mrs. Beck) on the right; a little beyond which the road skirts the banks of the lake to its outlet near the bridge. From thence pass through the village of Sawrey, with LAKE FIELD (J. R. Ogden, Esq.) on the left, and return on its eastern side to Hawkshead.

From Hawkshead to the Ferry-house on Windermere, where there is a good and commodious inn, the road passes over hilly ground through the villages of Sawrey. The sight of Windermere from the top of the hill is extremely fine.

The tourist halting here for a while, ought, by all means, to visit the Station-house, which is within a short and pleasant walk of the Inn, and commands a beautiful prospect of nearly the whole extent of the lake. Proceed to Bowness by the Ferry, or, if there be an objection to crossing the Ferry, there is a good road, abounding in a delightful succession of changes, on the west side of the lake, 8 miles, to Ambleside.

KENDAL ROUTE.

Let us now go back to Lancaster, and conduct the stranger to WINDERMERE by way of Kendal.

LANCASTER,

The capital of the County Palatine of Lancaster, is very finely situated on a hill rising abruptly from the river Lune, which falls into the Bay of Morecambe at the distance of six miles. There is excellent accommodation at two good inns, the *King's Arms* and *Royal Oak*. On the summit of the hill is the Castle, a majestic structure originally built by Roger de Poictou in the 11th century, and re-edified by John of Gaunt, Duke of Lancaster, in the 14th. It has been greatly enlarged in modern times, and now serves as the county gaol. The Parish Church of St. Mary's, an ancient structure with a lofty tower, stands also on the Castle Hill. A handsome new church has been recently erected in Penny Street, and there are several other Episcopal and Dissenting Places of Worship in that town. The County Lunatic Asylum is a handsome building situated on Lancaster Moor, about a mile from the town, and is capable of accommodating 300 patients. The foreign commerce of Lancaster has been on the decline for many years, having been injured by the competition of Liverpool; and the river being difficult of navigation, in neap tides the larger ships generally unload at Glasson Dock, five miles distant from the town. Lancaster is connected with the principal towns of the county by a canal, which is carried over the Lune two miles from the town by a magnificent aqueduct, erected by the late Mr. Rennie. Lancaster is celebrated for the manufacture of mahogany furniture, and several cotton and silk mills have of late years been established here. The formation of the Railway to Lancaster has been of great importance to the town, and may be considered as the beginning of a new era in its history—transferring a listless and stationary community into one of those "hives of industry" by which the commercial character of this country is sustained. Market on Wednesday and Saturday.

From Lancaster to Kendal the distance, by railway, is accomplished in about an hour. The tourist, on arriving at Oxenholme Station, will have to change his carriage, and proceed by the Kendal and Windermere line (two miles) to KENDAL, where it may be worth his while to tarry for a short period; or, he may proceed at once to WINDERMERE.

KENDAL

Is the largest and most important town, though not the metropolis, of the County of Westmorland, and is situated principally on the west bank of the river Kent, in a pleasant and fertile valley encompassed by hills of considerable height. It consists of two main streets, in continuity, from north to south, from which all the other streets, lanes, and alleys branch off at right angles. Excellent accommodations will be found at two good inns, *The King's Arms* and *Commercial Hotel.* Kendal is a place of great antiquity, but the re-erections and enlargements give it a modern appearance. The houses are built of mountain limestone, peculiarly rich in organic remains, which is obtained in great abundance from Kendal Fell, on the west side of the town. This material is quarried out in large blocks, and, being capable of a very high polish, is also extensively used in the manufacture of chimney-pieces. The woollen manufactures of this kingdom were first established, by act of Parliament, in Kendal. John Kemp, a manufacturer from Flanders, was the person who first received "protection" to establish himself in this country, and he settled here in the reign of Edward III. (1331). To the woollen manufacture this town has long been indebted for its prosperity; latterly, however, owing to competition in Yorkshire, &c., the trade in coarse woollens has not increased, and some of the manufacturers have turned their attention to the manufacture of carpets and worsted goods. Kendal is noted for its Railway Wrappers, and also for its Carpets, both of which obtained medals at the Great Exhibition.

The Castle stands upon a verdant knoll of oval shape on the east side of the town, and commands a pleasing and extensive prospect to the north and south-west. This fortress was the seat of the ancient Barons of Kendal, and the birth-place of Catherine Parr, the last wife of Henry VIII. No records have been preserved to establish the date of this castle. There is, however, very little doubt but it was raised altogether, or in part, by one of the first Barons of Kendal. If in part only by one of the first Barons, the completion of it must be assigned to those who lived in the 12th or early part of the 13th century. The circular tower of this castle is the most entire part of the ruin, and has evidently been the strongest; but the precise time when it was erected, and whether the rest of the building be coeval with it, must, it is to be feared, for ever remain in obscurity. The order of architecture and the arrangements of the apartments, however, bear an obvious resemblance to some of the castles (Cockermouth Castle for a particular instance) which have been referred to the time of the Conqueror. The date of the Castle's decay or destruction may fairly be taken from the attainder of Queen Catherine's brother, the Marquis of Northampton, in 1553, and as only nineteen years intervened between that event and the time that it has been proved to be in ruins (1565), the most plausible conclusion seems, that it was dismantled or thrown down in the Marquis's unsuccessful engagements against the

Crown, in favour of Lady Jane Grey. The Castle and part of the lands annexed to it have lately been purchased by William Thompson, Esq., M. P., Alderman of London.—For further particulars respecting the history of this venerable edifice, and the family of the Parrs, see the "*Annals of Kendal.*"

The Church, a Vicarage in the gift of Trinity College, Cambridge, is a spacious five-aisled Gothic structure, and has been recently renovated at a considerable expense. It now forms one of the most elegant structures, internally, in the North of England. In it are three "quires," or private chapels, memorials of the ancient dignity of three neighbouring families, the Bellinghams, Stricklands, and Parrs.

The Natural History Society's Museum is worthy the notice of passing Visitors. A considerable collection of specimens will be found in the following branches of natural science—Mineralogy, Geology, Ornithology, Botany, &c. Admission, gratis, on obtaining a ticket from a Subscriber.

There are many pleasant walks in the vicinity of Kendal, and to those who feel an interest in Botanical and Geological pursuits, this neighbourhood has peculiar attractions. The WALK TO SCOUT SCAR, a noble limestone cliff about two miles to the west of Kendal, is especially interesting. The Naturalist who may wander to this beautiful spot will find abundant material for interesting examination. For the use of the Botanist a list of the rarer plants in this locality, as well as of the land shells, will be given at the end of the volume. Many of the less common species of land shells, especially of the Helix, Pupa, and Vertigo genus, will be found in their peculiar habitats in the course of a ramble across the face of the hill. Several of the beds of the (carboniferous) limestone, exposed in the escarpment, yield in abundance the characteristic shells and corals of this formation. Part of the upper Ludlow rocks of the Silurian system may be seen cropping out beneath the limestone, and rising through the peat-moss, in rounded masses, in various parts of the valley below. A walk round the southern extremity of the fell, by the new road down to the village of Brigsteer, will amply repay the Geologist by a beautiful section through the limestone and Silurian beds, down to the level of the moss, which is exposed there. We may observe, that the most characteristic fossils of the neighbourhood may generally be purchased from Collectors in Kendal; and the collection of Mr. John Ruthven, an excellent practical geologist, in Castle-street, is especially deserving of remark. The travelled blocks of greenstone, &c., from the lake rocks, resting on different parts of the fell, and in many instances crowning its highest elevations (*blocs perchès*), will not be passed unnoticed.

Excursions from Kendal.

SHAP WELLS.—A spacious Hotel with Baths and every accommodation for Visitors has been erected at this place, Shap Spa is stated by Mr. Alderson, in his "Treatise," to be a most genial and santive saline spring, milder than the Harrowgate, and more active than the Gilsland Water, and in its properties nearly allied to that of Leamington. It is much frequented by persons seeking health or recreation. The distance from Kendal is 15 miles, and the Lancaster and Carlisle Railway passes within a few hundred yards of the Hotel.

HAWES WATER.

To HAWES WATER, through LONG SLEDDALE.

4½	Watch Gate	4½	2	Sadgill Bridge	9½
3	Long Sleddale Chapel ...	7½	4½	Chapel Hill	14

"Following the road from Kendal to Shap for about four miles, the traveller will see a deep and narrow valley, turning somewhat westward into the mountains: this is Long Sleddale, into which a cross road down a steep hill will conduct him. If not one of the grandest character, it has the advantage at least of being thoroughly free from the intrusion of art. There is nothing to mar its harmony: and while passing along the narrow lanes, enclosed by thickly-lichened walls, tufted with wild flowers and crested by hedges, as the eye rests on the brilliant green of the meadows, the sparkling purity of the stream, or the autumnal tints of the copses, we heartily rejoice in our emancipation from the turnpike-road, and acknowledge this to be a genuine and lovely specimen of pastoral scenery. The upper portion of the dale is bleak and sterile, and the ascent to the summit of the pass which divides it from Mardale is wearisome; but on attaining the summit, the bird's-eye view of the deep green secluded glen beneath, and the abruptness and ruggedness of the descent, will strike one who is unaccustomed to mountain-passes with surprise and delight. There is a small public-house, the *White Bull*, where rough but clean accommodation may be had, at Mardale Green, about a mile above the head of Hawes Water. The lake is three miles long,—'a sort of lesser Ullswater,' Mr. Wordsworth says, 'with this advantage, that it remains undefiled by the intrusion of bad taste; and, from the remotness of the situation, it is long likely to remain so. The eastern bank is clothed by natural wood, of no great size or beauty, but richly feathering the hill side and shore of the lake."

The Tourist may return to Kendal through Kentmere, or, he may proceed to Bowness by striking across the summit of High Street on the right from the pass of Nanbield, and descending into the valley of Troutbeck, which opens to Windermere a little below Low Wood. The distance from Mardale to the public-house at Troutbeck is about six miles, from thence to Bowness it is four miles. High Street is 2,700 feet above the level of the sea. Remains of the Roman road from Kendal (*Concangium*) to Penrith (*Petriana*), may be traced along its summit. The views from it are extremely fine, and the road all the way to Bowness abounds in charming prospects.

From KENDAL to WINDERMERE, the distance, by railway, is nine miles.

WINDERMERE

Is a small post-town, rapidly rising into importance, at the terminus of the line. Here the tourist will find excellent accommodation at the *Windermere Hotel*, a commodious establishment, under the superior management of Mr. and Mrs. Rigg, which

> "Overlooks the bed of Windermere
> Like a vast river, stretching in the sun.
> With exultation at his feet he sees
> Lake, islands, promontories, gleaming bays,
> A universe of Nature's fairest forms
> Proudly revealed with instantaneous burst,
> Magnificent, and beautiful, and gay."

Before the introduction of the railway into this district, there was not a house on the spot that now forms the site of this flourishing village. The Church, a neat edifice, a few hundred yards on the Ambleside road, was built by the Rev. J. Addison, the present incumbent, and was afterwards enlarged, by the addition of one aisle, at the expense of John Braithwaite, Esq., of Orrest-Head. A College, designed chiefly for the education of the Sons of the Clergy, is about to be erected at Windermere, and lodging-houses and villas are springing up in every direction. Private lodgings may be had in the village, if required; and strangers will find a good Circulating Library at Mr. J. Garnett's stationer and postmaster.

There are numerous pleasant walks in the neighbourhood, which will be readily pointed out to tourists, and from the top of the hill behind the hotel a magnificent view of the surrounding country may be obtained. Our outline Diagram of the Mountains, as seen from Low Wood, will assist the stranger to gain a knowledge of their names.

There is an interesting walk, abounding with rich and varied scenery, along a public footpath through the woods above ELLERAY, formerly the residence, and until lately the property, of Professor Wilson. The present proprietor, Mr. Eastted, has formed extensive drives through the Elleray estate, which are, of course, private; but, through the liberality of the owner, the public are permitted access, under certain restrictions, by tickets, which can be obtained *only* at the post-office.

There is also an agreeable walk through the copses in the direction of the lake, by an ancient bridle-road, which is entered through a gate at the Parsonage (the second below the Church). Immediately on entering this gate, the road turns into a field on the left hand, passes in front of the house, and soon enters the wood. This road comes out into the lane leading from Cook's House to Bowness, at the farm called Miller Ground. On rising the hill on the right, the tourist will soon join the Ambleside road, having on the left, overlooking the woods of Calgarth, a view of Windermere, with the Pikes of Langdale, forming a landscape of surpassing richness.

The beautiful valley of Troutbeck may be conveniently visited from Windermere, and the ascent of High Street, at its head, is more easily accomplished from this than from any other point.

TROUTBECK.

Tourists visiting Troutbeck on horseback or in carriages will have to proceed on the Ambleside road for about a mile, and turn to the right at Cook's House. Pedestrians may take a short cut through the Elleray woods, by the public foot-path, which joins the Troutbeck road at St. Catherine's (Lord Bradford's.) This road leads straight into the valley; but, before reaching the chapel, parties should take a lane to the left, through the village, which is somewhat remarkable for its cottage architecture, and more favourable for seeing the beauties of the vale. In the village is a small public-house, called the "*Mortal Man*," which name it acquired from the following humourous distich, inscribed upon a sign-board which formerly hung over the door:—

> "Oh! mortal man, that liv'st on bread,
> How comes thy nose to be so red?
> Thou silly ass, that looks so pale,
> It is by drinking BIRKETT's ale!"

This sign-board, depicting the portraits of two well-known characters in the vale—one of them rubicund and jolly, with a nose giving unmistakeable evidence of a love of the bottle, the other with a visage remarkable for the longitude of its outline and its cadaverous hue—was painted by a clever and eccentric artist of the name of Julius Cæsar Ibbotson, who resided in Troutbeck about thirty or forty years ago, and who was probably also the author of the above lines. Troutbeck now mourns the loss of this noted sign, which was some time ago removed to Allithwaite, near Cartmel, by the landlord, whose property it was, and where, by long exposure to the weather, both the picture and poetry were obliterated, and thus lost to the world.

From the inn to the head of the valley the distance is about three miles, the road skirting the hill on the western side of the vale, and abounding in scenes of great pastoral beauty. The mountains on the north-east are those of Kentmere, namely, the Yoke, Hill Bell, Froswick, and High Street, which closes in the valley at its head. If the high road were pursued, the tourist would be led to Kirkstone and Patterdale.

Troutbeck was the birth-place of the father of Hogarth, the greatest of our dramatic English Painters. The paternal uncle of the painter, Thomas Hogarth, better known whilst living by the familiar name of "Auld Hoggart," flourished in this vale about a century and a half ago. He was a rustic poet and satirist, "whose rude and witty productions (in the opinion of Adam Walker, the naturalist, also a denizen of this valley,) reformed the manners of the people as much, at least, as the services of the clergyman." An old manuscript volume of his poetry, evincing considerable skill at versification, has lately been discovered. The pieces consist, principally, of short dramas, in verse, the interest arising from the incidents of low rustic intrigue, songs, epigrams, &c., &c., from which a selection has recently been made and published in a pamphlet, together with some account of his life and eccentricities.

BOWNESS.

BOWNESS is about a mile and a half from the village of Windermere, and is situate

> "Midway on long Winander's eastern shore,
> Within the crescent of a pleasant bay."

It is favourable for aquatic excursions, both by the steamers, which pass and repass several times in the course of a day, and also by pleasure-boats, which are kept and let out to parties desirous of enjoying the scenery of the lake from its surface. Bowness contains two comfortable and commodious hotels, the *Royal*, Bowness's, (late Ullock's), so designated since the visit of the late Queen Dowager, and the *Crown* (Cloudesdale's). The Church is an ancient structure with a square tower, dedicated to St. Martin. The chancel window is of painted glass, and was brought hither from Furness Abbey after the destruction of that monastery. (See p. 6, for a description of this window.) The remains of the late learned Bishop Watson, of Llandaff, rest in the Church-yard, close by the eastern window. His tomb bears the following simple and unpretending inscription; "Ricardi Watson, Episcopi Landavensis, cineribus sacrum, obiit Julii 1, 1816, Ætatis 79." A handsome school-house looks down from an eminence in the centre of the village, and stands as a monument of the munificence of the late John Bolton, Esq., of Storrs Hall, who erected the edifice at his own expense.

WINDERMERE LAKE.

Windermere is the largest of the English Lakes, being ten miles in length, and more than a mile at its greatest breadth. Its two principal feeders are the rivers Brathay and Rothay, which join near Croft Lodge, and pour their united waters into the Lake. The Brathay rises in the group of lofty mountains between Langdale and Borrowdale. The Rothay issues partly from Rydal Water and partly out of the hills at the head of Ambleside. A circumstance very interesting to the Naturalist should be mentioned here. The *Char* and *Trout*, at the approach of the spawning season, may be seen proceeding together out of the lake up the stream to the point where the Brathay and Rothay meet, when they uniformly separate, as if by mutual arrangement, the char always, and all of them, taking the Brathay, and the trout the other stream, the Rothay. Is it a difference in the quality of the waters, or some geological peculiarity in the river beds, that influences these fish in their choice of streams?

The lower part of Windermere is now, from the facilities afforded by the Steamers, more frequently visited than formerly. It has many interesting points of view, especially at Storrs Hall and at Fellfoot, where the Coniston mountains peer nobly over the western barrier, which elsewhere, along the whole lake, is comparatively tame. For one also who has ascended the hill from Graythwaite on the western side, the Promontory called Rawlinson's Nab, Storr's Hall, and the Troutbeck Mountains, about sun-set, make a splendid landscape. The view from the Pleasure-house of the Station near the Ferry has suffered much from larch plantations, this mischief, however, is gradually disappearing, and the larches, under the management of Mr. Curwen, are giving way to the native wood. Windermere ought to be seen both from its shores and from its surface. None of the other lakes unfold so many fresh beauties to him who sails upon them. This is owing to its greater size, to the islands, and its having *two* vales at the head, with their accompanying mountains of nearly equal dignity. Nor can the grandeur of these two terminations be seen at once from any point, except from the bosom of the lake. The Islands may be explored at any time of the day; but one bright unruffled evening, must, if possible, be set apart for the splendour, the stillness, and solemnity of a three hours' voyage upon the higher division of the lake, not omitting, towards the end of the excursion, to quit the expanse of water, and peep into the calm river at its head, which, in its quiet character,* at such a time, appears rather like an overflow of the peaceful lake itself, than to have any more immediate connection with the rough mountains whence it has descended, or the turbulent torrents by which it is supplied. Many persons content themselves with what they see of Windermere during their progress in a boat from Bowness to the head of the lake, walking thence to Ambleside. But the whole road from Bowness is rich in diversity of pleasing or grand scenery; there is scarcely a field on the road side, which, if entered, would not give to the landscape some additional charm. In addition to the two vales at its head, Windermere communicates with two lateral Vallies; that of Troutbeck, distinguished by the mountains at its head; by picturesque remains of cottage architecture; and, towards the lower part, by bold foregrounds formed by the steep and winding banks of the river. This Vale, as before mentioned, may be most conveniently seen from Low Wood. The other lateral Valley, that of Hawkshead, is visited to most advantage, and most conveniently, from Bowness; crossing the lake, by the ferry—then pass the villages of Sawrey, and, on quitting the latter, you have a fine view of the Lake of Esthwaite, and the cone of one of the Langdale Pikes in the distance.

Numerous Islands adorn the surface of this lovely lake, the largest of which, BELLE ISLE, the summer residence of H. Curwen, Esq., contains upwards of thirty acres. This island is well wooded, and being intersected by shady walks, open to tourists, affords a pleasent change to those who land upon its shores. LADY HOLME, a small island nearly opposite to Rayrigg, had in the time of Henry VIII. a chapel dedicated to our Lady within its small territory, belonging to Furness Abbey, but no traces of this sanctuary are left to mark its site.

Many pleasant walks will be found in the neighbourhood of Bowness; and one to the top of Bisket How, a small eminence overlooking the valley, affords extensive views of the surrounding country.

* Since this was first written, the natural beauty of this scene has been grievously impaired.

The Troutbeck Excursion may be made conveniently from Bowness, but this Station is too remote from the adjacent mountains for excursions, which should be taken from Low Wood or Ambleside.

The road from Bowness to Ambleside is partly through wooded ground, passing RAYRIGG, the residence of Major Jacobs, on the left, on a slight elevation above the surface of the lake, at an agreeable distance from the road. On rising the hill above Rayrigg, it passes Millar Ground, an ancient farm-house, and soon joins the Ambleside road at Cook's House, before mentioned.

The road from this point to Ambleside passes Troutbeck Bridge about a mile distant, where a neat residence, called IBBOTSHOLME, (S. Taylor, Esq.,) has lately been erected. CALGARTH PARK (T. Swinburn, Esq.), formerly the seat of the learned and venerable Bishop Watson, of Llandaff, is on the left. Also, on the left, a little further on, is ECCLERIGG, the residence of Luther Watson, Esq., and on the right HOLBECK COTTAGE (Miss Meyer). —Presently, the tourist will reach

LOW WOOD INN,

A mile from the head of Windermere. This is a most pleasant halting-place; no inn in the whole district is so agreeably situated for water-views and excursions; and the fields above it, and the lane that leads to Troutbeck, near it, present beautiful views towards each extremity of the lake. From this place, and from Ambleside, rides may be taken in numerous directions, and the interesting walks are inexhaustible; a few of these will hereafter be particularized. The road from Low Wood to Ambleside, a distance of two miles, passes DOVE NEST, for a short time in the summer of 1830 the favourite retreat of the late Mrs. Hemans, and WANSFELL HOLM, the the seat of the Rev. — Hornby, Rector of Winwick, from whence, across the head of the lake, at the foot of Loughrigg Fell, is seen CROFT LODGE, the residence of J. Holme, Esq., of Liverpool. From this point, also, looking in the same direction, the picturesqne Chapel of Brathay, at the entrance of the vale of Langdale, is visible. This Chapel is in the Italian or Swiss style of architecture, and was built by Giles

Plate 1.

Mountains as seen from Low Wood Inn.

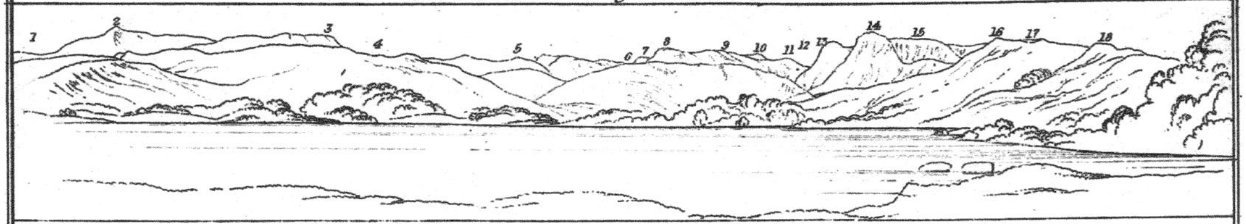

1 Walnay Scar
2 Coniston Old Man
3 Wetherlam
4 Skelwith
5 Crinkle Crags
Pike of Bliscow (below)
6 Lingmoor
7 Scawfell Pike
8 Bowfell
9 Hanging Knotts
10 Great End
11 Glaramara
12 The Stake Pass
13 Pike of Stickle } Langdale Pikes.
14 Harrisons Stickle }
15 Paveyark
16 Loughrigg Fell
17 Easedale Head
18 Lowrigg Fell

Mountains as seen at an elevation on the road opposite the Inn at Strandsin Wasdale.

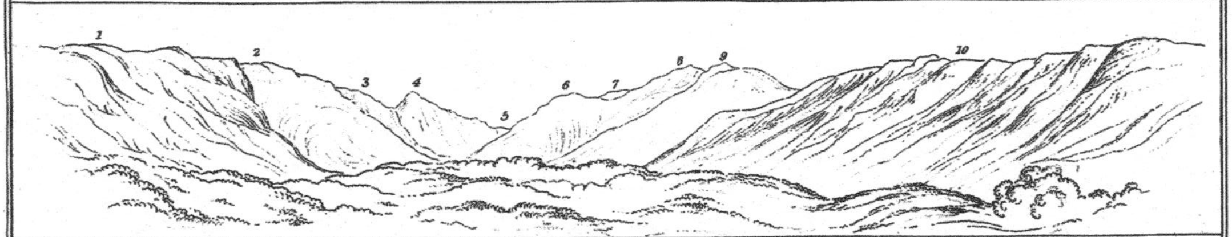

1 Buckbarrow Pike
2 Middlefell
3 Yewbarrow
4 Great Gable
5 Sty Head Pass
6 Lingmell
7 Great-end Crag
8 Scawfell Pike
9 Scawfell
10 Screes

J. Flintoft delt. W.H. Lizars sc. Edinr.

Redmayne, Esq., of London, whose summer residence, **Brathay Hall**, is seen a little to the south. The Whitehaven mail and other coaches pass daily through Ambleside, leaving the Windermere Station on the arrival of the trains.

Excursions from Low Wood.

From this Inn, which has lately been much enlarged, the following Excursions may be made, and may be taken also with the same convenience from Ambleside.

WALK to SKELGILL from LOW WOOD.

1¼	Low Fold	1¼	½	Troutbeck road	3½
1½	Skelgill	2¾	1	Low Wood	4½
¼	Low Skelgill	3			

CIRCUIT from LOW WOOD by AMBLESIDE, KIRKSTONE, and TROUTBECK.

1¾	Ambleside	1¾	4¼	Troutbeck	10
4	Guide-post on Kirkstone ...	5¾	2	Low Wood	12

WALK or HORSE-RIDE through TROUTBECK and APPLETHWAITE to BOWNESS, or back to LOW WOOD.

2	Guide-post in Troutbeck ...	2	2½	Cook's House	5¼
¾	The How, in Applethwaite	2¾	2	Bowness	7¼

If the return is from Cook's House to Low Wood, the round will be 8 miles.

These Excursions abound in delightful prospects, and the view from the top of the hill about a mile from the inn, on the Troutbeck road, is the finest of its kind amongst the Lakes. From this point the islands of Windermere are seen "almost all lying together in a cluster, below which all is loveliness and beauty—above, all majesty and grandeur."

Excursions from Ambleside.

Ambleside is a small market town, situate in the Vale of the Rothay, one mile north of Windermere. Good accommodations are here provided for Tourists at the *Salutation Hotel* (Donaldson), the *Commercial Inn*, (Armer), and the *White Lion*, (Townson), as well as at private lodgings; and, as the town is in the neighbourhood of many very interesting excursions, Visitors to the Lakes usually make it their head-quarters for some time.

A handsome church has recently been built here by subscription, and forms a conspicuous and pleasing object in the vale. There are two Circulating Libraries in the town—one at the Post-Office, and the other kept by T. Troughton, the Parish Clerk; and here also has recently been established a Branch of the Kendal Bank, under the management of Mr. Newby, draper, in that part of the market-place called Cheapside. Ambleside was formerly a Roman station (the *Dictis* of the Notitia), and some slight traces of a fortress are perceptible in a field at the head of Windermere, where fragments of tesselated pavement, urns, and other Roman relics have been dug up. This station was established, undoubtedly, as a check upon the pass of Kirkstone, Dunmail-raise, and of Hardknott and Wrynose.

VALES OF GREAT AND LITTLE LANGDALE.

1	Clappersgate	1	1¼	Lisle Bridge	11
1¼	Guide Post	2¼	2	Langdale Chapel	13
¾	Skelwith Fold	3	1½	High Close	14½
1	Colwith Bridge	4	½	First sight of Grasmere	15
1½	Little Langdale Tarn	5½	2	Grasmere	17
2¼	Blea Tarn	7¾	4	Ambleside	21
2	Wall End	9¾			

This is a charming excursion. From Ambleside go to Clappersgate, where cross the Brathay, and proceed, with the river on the right and the chapel on the left hand, to the hamlet of Skelwith-fold. When the houses are passed, turn, before you descend the hill, through a gate on the right, and from a rocky point is a fine view of the Brathay river, Langdale Pikes, &c.; thence to Colwith-force; and, after passing through a gate, a short distance from Little Langdale Tarn, the ancient road from Kendal to Whitehaven takes the left hand; the one to be pursued turns to the right, leading over the common to Blea Tarn. The scene in which this small piece of water lies, suggested to the author the following description (given in his Poem of the Excursion), supposing the spectator to look down upon it, not from the road, but from one of its elevated sides.

"Behold!
Beneath our feet a little lowly Vale,
A lowly Vale, and yet uplifted high
Among the mountains; even as if the spot
Had been, from earliest time, by wish of theirs,
So placed, to be shut out from all the world!

Urn-like it was in shape, deep as an Urn;
With rocks encompassed, save that to the South
Was one small opening, where a heath-clad ridge
Supplied a boundary less abrupt and close;
A quiet treeless nook,* with two green fields,
A liquid pool that glittered in the sun
And one bare Dwelling; one Abode, no more!
It seemed the home of poverty and toil,
Though not of want: the little fields, made green
By husbandry of many thrifty years,
Paid cheerful tribute to the moorland House.
—There crows the Cock, single in his domain:
The small birds find in spring no thicket there
To shroud them: only from the neighbouring Vales
The Cuckoo, straggling up to the hill tops,
Shouteth faint tidings of some gladder place."

At this point the Langdale Pikes, with Gimmer Crag between, rising from the unseen vale below, appear in a new and noble aspect; indeed, a more dignified and impressive assemblage of mountain lines scarcely exists in the North of England. The highest Pike, called Harrison Stickle, is perhaps about three miles from the eye, but Stickle Pike, receding towards the Pass of the Stake into Borrowdale, is more than four. After leaving the Tarn, the road descends rapidly to Wall End, at the head of Great Langdale,† from whence it is recommended to proceed to Millbeck, a farm-house across the meadows, a mile distant, and see

DUNGEON GILL.—The Gill, having its source between the Pikes, passes through a deep cleft in the mountain, into the cheeks of which a rock from the neighbouring heights hath fallen, and got so wedged in as to form a grotesque natural arch,

——"a spot which you may see
If ever you to Langdale go;
Into a chasm a mighty block
Hath fallen, and made a bridge of rock:
The gulf is deep below;
And, in a basin black and small,
Receives a lofty waterfall."

* No longer strictly applicable, on account of recent plantations.

† The upper portion of the Vale of Langdale, which lies in the direction of the valley which stretches westward towards Bowfell and Crinkle Crags, bears the name of OXENDALE.

LANGDALE PIKES may be conveniently ascended from Mill-beck, where a guide may be obtained. The best ascent is by a peat road to STICKLE TARN, a pretty circular piece of water, celebrated for its fine trout, reposing under the steep rocks of Pavey Ark, and thence to the top of the Pike called Harrison Stickle, which is 2,400 feet in height. Although this Pike is inferior in elevation to many of the neighbouring mountains, the views from it are interesting and extensive, especially in looking over the Vale of Great Langdale, towards Windermere, and over the open country to the south and south-east. From Stickle Pike, which rises like a cone a little to the north, there is a fine view of Skiddaw and the Vale of Bassenthwaite, the former of which is seen but partially, and the latter not at all, from Harrison Stickle. Great Gable rears his head to the west, Great End is a little nearer the eye, and Scawfell and the Pikes are seen pre-eminent over the summit of Bowfell. Crinkle Crags are a continuation of Bowfell on the south, and to the south-west, looking over the lonely valley of Little Langdale, are the Coniston mountains: on the east are the mountains of Rydal and Grasmere, and on the north-east the Helvellyn range forms a prominent feature in the landscape. In the north, Saddleback in the distance, presents his front to the spectator.

On leaving the Pikes, follow the road down Great Langdale, as far as the Chapel, passing Thrang Crag Slate Quarry on the left, which those who take an interest in geological science ought not to omit looking at. Near the Chapel there is a small ale-house, from which it is five miles to Ambleside. The road is either by Loughrigg Tarn, or by Rydal and Grasmere waters. The latter course is much to be preferred. The road strikes off on the left, near the Chapel, and in winding up the hill the whole Vale of Langdale, with the small Lake of Elterwater and Loughrigg Tarn, are seen to advantage. The view from High Close is exquisite, and Mr. Green says, "there is not a finer thing in Westmorland." A few hundred yards from this point will bring you in sight of the Lake and Vale of Grasmere, from whence, turning southward, it is four miles on the main road to Ambleside. This excursion is altogether twenty-one miles (if Dungeon Gill and the Pikes are visited), of which, though assisted by a carriage, it will be necessary to walk from five to seven miles.

Stock Gill Force, half a mile from Ambleside, is a most interesting Waterfall, if seen to advantage, but its beauties are in a great degree lost to the generality of visitors, who see the fall only from the footpath skirting the top of the bank, and almost perpendicularly from the bottom of the channel. The spectator looks down upon the scene rather than upwards or horizontally, and his view of the water is likewise impeded by a redundancy of wood. Stock Gill rises in the Screes, on the side of Scandale fell, not far from Kirkstone, and, passing through Ambleside, joins the river Rothay a quarter of a mile below the town, about four miles from its source. This rivulet is among the finest of its kind in the Lake District. The way to the Waterfall is through the stable-yard of the Salutation hotel..

AQUATIC EXCURSIONS ON WINDERMERE.

¾	Landing at Waterhead	¾
¾	Mouth of the river	1½
1	Pull Wyke	2½
1¾	Low Wood Inn	4¼
½	Holme Point	4¾
½	Return to the mouth of the river	5¼
¾	Landing	6
¾	Ambleside	6¾

To the Landing at Waterhead, where boats are moored, the walk is three-quarters of a mile. After taking boat, steer a short and attractive course by skirting the deeply-indented coast of Brathay into Pull Wyke, a pretty bay surrounded by rich woods, over which peep the Loughrigg and other elevated summits; and from Pull Wyke proceed by the grounds at Low Wray to the craggy and wooded promontory a little southward. From this place make for the Inn at Low Wood in a direct line, and see the Langdale and Rydal Mountains in two several and distinct arrangements, separated by the imposing heights of Loughrigg. Then return to the mouth of the Brathay by Holm Point, and up the river to the landing place.

From AMBLESIDE to the FERRY, by Water.

1½	Mouth of the river by the Landing	1½
3	Belle Grange	4½
2½	Ferry-house, passing between the Lily of the Valley Holmes	7
¾	From the Ferry-house to the Landing on Curwen's Island	7¾
1¾	Round the Island	9½
4½	From the Pier to the Head of the Lake	14

The best situation on the water for a view of the country around is about half a mile from the junction of the Brathay with the lake, and parties in an excursion downwards will do well to pass in that direction, and from that point rather near to the Lancashire shore, by which the high lands at Rydal, Ambleside, Troutbeck, and Applethwaite, will be seen to the greatest advantage, particularly Hill Bell and the neighbouring summits. In proceeding towards the Ferry, that part of the lake between the two islands, called the Lily of the Valley Holmes, having the Station-house about a mile from the eye, and as side-screens the bold and wooded elevation above Harrow Slack on the right, and Curwen's Island on the left, forms a lovely picture. Rather than first touch at the great Island, it will be better to row direct for the Ferry-house, thence walk to the Station, and afterwards return to the Ferry. From the Ferry, Curwen's Island should be visited, after leaving which the party may visit Bowness, or return direct to Ambleside.

By a tourist halting a few days in Ambleside, the *Nook* also might be visited—a spot where there is a bridge over Scandale-beck, which makes a pretty subject for the pencil. And, for residents of a week or so at Ambleside, there are delightful rambles over every part of Loughrigg Fell and among the enclosures on its sides; particularly about Loughrigg Tarn, and on its eastern side about Fox How and the properties adjoining to the northwards.* A few out of the main road are particularized in the following Tables:—

* Sergt. (now Judge) Talfourd, in his "Vacation Rambles," speaking of Loughrigg Fell, says, "This beautiful piece of upland might seem a platform—if such a phrase did not belie its waving, rock-ribbed, and pinnacled surface—built by Nature, to enable her true lovers to enjoy, in quick succession, the most splendid variety she can exhibit. On one side, from the gently ascending path, bordered by scanty heather, you embrace the broader portion of Windermere, spreading out its arms as if to embrace the low and lovely hills that unfold it—a view without an angle or a contrast—a scene of perfect harmony and peace, Ascend a lofty slab of rock, not many paces onward, and you have lying before you the delicious vale of the Rotha—a stream gliding through the greenest meadows—with Fairfield beyond, expanding its huge arms as of a giant's chair, and with Fox How in the midst, where the great and good Dr. Arnold—great in goodness—embraced the glories of the external world, with all the earnestness of his generous and simple nature, and nourished that sense of the imaginative and harmonious aspects of humanity and faith which grew clearer and deeper as he advanced in years. Wind your way through two small valleys, each having its own oval basin, and from another height you may look down on the still mirror of Rydalmere, with its small central island, the nest of herons,

From AMBLESIDE, under Loughrigg Fell, to GRASMERE.

½	Rothay Bridge	½	¾	Dale End	4¾
1½	Pelter Bridge (leave on the right)	2	¾	Grasmere Church	5½
¼	Coat How	2¼	4	Ambleside	9½
1¾	Red Bank	4			

This is one of the finest Walks in the country. The tourist must take the road to Clappersgate, and, after crossing Rothay Bridge, enter a gate on the right hand. He will pass in regular succession Millar Bridge Cottage on the left; Fox How (Mrs. Arnold) on the right; Fox Ghyll (H. Roughsedge, Esq) on the left; Loughrigg Holme (Misses Quillinan); Spring Cottage (Wm. Peel, Esq.); Ebenezer Cottage; and Field Foot (W. D. Crewdson, Esq.), also on the left. Rydal Hall, the seat of Lady le Fleming, standing in an extensive park, richly adorned with numerous stately forest trees, and Rydal Mount (Mrs. Wordsworth), are prominent objects from several parts along the road; and the mountains of Rydal Head, Fairfield, and Nab Scar on the north-east, and Loughrigg fell on the western side of the valley, present many fine combinations of scenery. On reaching Pelter Bridge he must leave it on the right, taking the road by Coat How; and on arriving at the top of the lane he will come in view of Rydal Lake. He must keep the high terrace road, which leads to Red Bank, and forward to Grasmere, from whence he may return to Ambleside by the Keswick road. This walk may be curtailed on arriving at Pelter Bridge, before named, by crossing it, and returning through Rydal to Ambleside.

LOUGHRIGG TARN and GRASMERE.

1	Clappersgate	1	¼	The Oaks	3
1¼	Guide-post	2¼	3	Grasmere Church	6
½	Loughrigg Fold	2¾	4	Ambleside	10

and following the valley to Grasmere with its low white church-tower, beyond the figured crest of Helm Crag, behold the vast triangle of Skiddaw filling the distance; while midway, just rising above green mountains, you may see the topmost *rind* of Helvellyn, curved in air, with one black descent just indicated; and, when the eye has been satiated with loveliness, look down just below on a mansion at the foot of Nab Scar, the dwelling of the Poet, not of these only, but of all earth's scenes; who, disdaining the frequent description of particular combinations of its beauties, has unveiled the sources of profoundest sentiment they contain; and, more than any writer who ever lived, has diffused that love of external nature which now sheds its purifying influence abroad among our people. Pass from thence to the highest point of all this region, and look down, beyond the calm round tarn of Loughrigg, into a magnificent chaos, the Langdale vales, with the ribbed pike of Scawfell beyond them, and in the midst those Pikes, which, yielding to many of the surrounding hills in height, surpass them all in form."

"LOUGHRIGG TARN," says Wordsworth, "resembles, though much smaller in compass, the Lake Nemi, or *Speculum Dianæ*, as it is often called, not only in its clear waters and circular form, and the beauty immediately surrounding it, but also as being overlooked by the eminence of Langdale Pikes as Lake Nemi is by that of Monte Calvo."

ELTERWATER.

The foot of Elterwater, either by Skelwith Bridge or Loughrigg Fold, over Little Loughrigg, is 3½ miles from Ambleside. Extensive Gunpowder Works are carried on at Elterwater.

WANSFELL PIKE.

1	Low Fold, and along a Terrace Road under Strawberry Bank	1	1 Wansfell Pike	3
1	Skelgill	2	1 Waterfall Lane	4
			1 Ambleside	5

RYDAL WATERFALLS.

1½	Lower Fall	1½	2 Ambleside	4
½	Higher Fall	2		

These two pretty water-falls are pointed out to every one, and may be seen on application at the Cottage near Rydal Chapel. The upper fall is in a glen above the Hall, but the lower fall, which is the more beautiful, is seen from a summer-house in the pleasure-grounds, and is thus described in one of Mr. Wordsworth's earliest poems:—

> "—— With sparkling foam, a small cascade
> Illumines from within the leafy shade,
> While thick above the rills, the branches close,
> In rocky basin its wild waves repose:
> Beyond . . . .
> The eye reposes on a secret bridge,
> Half grey, half shagged with ivy to its ridge.
> There bending o'er the stream, the listless swain
> Lingers behind his disappearing wain."

FAIRFIELD is the high mountain closing on the north the domain of Rydal, with an elevation of 2950 feet.—Commence the ascent to Fairfield at Rydal by the road between Rydal Hall and Rydal Mount, beyond which there is a green lane that leads to the Common, whence it is a steep and craggy climb to Nab Scar. From a certain point on Nab Scar there is an exquisite view commanding eight lakes: viz. Windermere, Blelham Tarn, Esthwaite Water, Rydal Water, Coniston Water, Elterwater,

Grasmere Lake, and Easedale Tarn. The traveller, if so inclined, may proceed to the top of Fairfield by following the ridge; and return to Ambleside by Nook End Bridge, over the High and Low Pikes. The distance is about ten miles.

From AMBLESIDE to HAWES WATER, over High Street

3	Woundale	3	2	Junction of High Street with Riggendale; Blea Water on the right	8
3	By Troutbeck Tongue to High Street, where Hays Water is seen on the left	6	2	Chapel Hill	10

To HAWES WATER, through Troutbeck and Kentmere.

4	Troutbeck	4	4	Nanbield	11
3	Kentmere Church	7	2	Chapel Hill	13

HAWES WATER does not exceed three miles in length, and varies in width from half a mile to a quarter. It is seldom visited by tourists, though the solemn grandeur of its rocks and mountains is exceedingly impressive. (See p. 21.)

From AMBLESIDE to HAYSWATER.

7	Low Hartshope	7	2	Return by Low Hartshope	11
2	Hayswater Head	9	7	Ambleside	18

From AMBLESIDE to ANGLE TARN.

7	Low Hartshope	7	1	Low Hartshope	9
1	Angle Tarn	8	7	Ambleside	16

HAYS WATER and ANGLE TARN are situated on the west side of High Street, and are celebrated for the fine trout with which they abound.

YEWDALE.

3	Skelwith Bridge	3	2	Shepherd's Bridge	7
1	Turn on the left at the top of the hill between Skelwith and Colwith Bridges	4	1	Black Bull Inn, Coniston ...	8
1	Oxen Fell	5	1	Water Head Inn	9
			8	Ambleside	17

TILBERTHWAITE.

7	Shepherd's Bridge, in Yewdale	7	5	Ambleside, over Colwith and Skelwith Bridges	15
1½	Tilberthwaite	8½			
1½	Little Langdale	10			

TILBERTHWAITE, returning by Elterwater Hall.

7	Shepherd's Bridge in Yewdale	7	2	Langdale Chapel, by Fletcher's Wood and Elterwater Hall	12
3	Little Langdale Road, by Tilberthwaite	10	5	Ambleside, by High Close, Grasmere, and Rydal Waters ...	17

From AMBLESIDE, round the Lake of WINDERMERE.

1	Brathay Bridge	1	7	Newby Bridge	15
4	High Wray	5	8	Bowness	23
3	Ferry House	8	6	Ambleside	29

From AMBLESIDE, round the Lake, by the Ferry Points.

1	Brathay Bridge	1	2	Bowness	10
7	Ferry House, by High Wray and Belle Grange	8	6	Ambleside	16

From AMBLESIDE by the Eastern Side of ESTHWAITE WATER and the Eastern Side of WINDERMERE.

5 Hawkshead	5	2 Bowness	11
2 Sawrey	7	6 Ambleside	17
2 Ferry-house	9		

AMBLESIDE TO PATTERDALE.

The distance from Ambleside to the Inn at Patterdale is ten miles, and the Pass of Kirkstone and the descent from it are very impressive; but this vale, nevertheless, like the others, loses much of its effects by being entered from the head; so that it is better to go from Keswick through Matterdale, and descend upon Gowborrow Park; you are thus brought at once upon a magnificent view of the two higher reaches of the lake. To such persons, however, as decide upon visiting Patterdale from Ambleside, the following information may be useful.—The road leaves Ambleside between the old Church and the Free Grammar School, and ascends gradually for upwards of three miles to the summit of the mountain pass on Kirkstone, where a small public-house has been recently erected, and is said to be the highest inhabited house in the kingdom. A large detached mass of rock, called, from its shape, Kirkstone, is seen on the left, near the top of the pass. On descending from Kirkstone, towards Patterdale, a new and interesting scene appears. Through a vista, you have a pretty peep at BROTHERSWATER and the heights of Patterdale in the distance. The road runs close to Brotherswater, and then turns at right angles across the meadows, where it meets with another road from Hartsop Hall at Cowbridge. Between Cowbridge and the Inn at Patterdale, the romantic valley of Deepdale runs up into the mountains on the left. At the right-angular turn of the road above mentioned, there is a bridle-road through the picturesque hamlet of Low Hartsop, along the side of Place Fell, which joins the main road again at Goldrill Bridge, a short distance from the Inn. The stream which flows through the hamlet of Low Hartsop, issues from the mountain tarn called HAYS WATER, situate on the western side of a ridge running up to High Street; and, in wet weather, the stream from Angle Tarn forms a pretty waterfall down the craggy side of Placefell.

The finest scenes on Ullswater lie between the Inn at Patterdale, and Lyulph's Tower, about four miles distant. The best way of seeing them is, to take a boat at the head of the lake, pass the islands called Cherry Holme and House Holme, and approach within sight of Stybarrow Crag. From House Holme, the views are exquisite in almost every direction. Proceed to LYULPH'S TOWER, an erection built by the late Duke of Norfolk for a pleasure-house, now the property of Mr. Howard, of Greystoke. It stands a little above the road in a part of Gowbarrow Park, and from the front of it are seen fine views of the lake. From Lyulph's Tower, a guide to ARA FORCE, about a quarter of a mile distant, may always be had. In returning to the Inn, it is advisable to row across the lake to a promontary at the foot of Placefell, and walk over the Point to Purse Bay, and thence by the farm of Blowick and Goldrill Bridge to the Inn. In this short walk, the magnificent scenery around the head of Ullswater is seen to the greatest advantage. See ULLSWATER.

After having duly explored the beauties of Ambleside and the neighbourhood, the next Station the tourist should aim at is KESWICK, which may be approached by various routes. The Direct Road is the only one that can be travelled over by carriages; but the hardy pedestrian might select from the several routes hereafter pointed out which he will pursue. There is however, a carriage road from Ambleside to Keswick by Wast Water, but the circuit is so extended that it is seldom adopted. This road is through Coniston, 8 miles—Broughton, 9 miles more—and over Birker Fell (a road somewhat rugged) by Santon Bridge to the Strands, near the foot of Wast Water, where there is a comforable Inn, 17 miles. From the Strands through Gosforth and Calder Bridge, thence over Coldfell to Lamplugh, and by Scale Hill to Keswick, 35 miles.—By Egremont, a better road, 37½ miles—making altogether a circuit of 69 miles.

AMBLESIDE to KESWICK, DIRECT.

1½	Rydal	1½	4	Smallthwaite Bridge	12¼
3½	Swan, Grasmere	5	3	Castlerigg	15¼
2	Dunmail Raise...	7	1	Keswick	16¼
1¼	Nag's Head, Wythburn ...	8¼			

A mile and a half from Ambleside the tourist reaches the romantic village of Rydal. On the right is seen, embosomed in wood, RYDAL HALL, the residence of Lady le Fleming, in whose grounds are the two pretty water-falls before mentioned.

RYDAL CHAPEL is a neat edifice, and will arrest the notice of the stranger on entering the village. It was erected and endowed at the expense of Lady le Fleming, to whom Mr. Wordsworth addressed the following lines lines on the foundation stone being laid:—

. . . .

O LADY! from a noble line
Of chieftains sprung, who stoutly bore
The spear, yet gave to works divine
A bounteous help in days of yore,
(As records mouldering in the Dell
Of Nightshade* haply yet may tell;)
Thee kindred aspirations moved
To build, within a vale beloved,
For Him upon whose high behests
All peace depends, all safety rests.

How fondly will the woods embrace
This daughter of thy pious care,
Lifting her front with modest grace
To make a fair recess more fair;
And to exalt the passing hour;
Or soothe it with a healing power
Drawn from the Sacrifice fulfilled,
Before this rugged soil was tilled,
Or human habitation rose
To interrupt the deep repose!

. . . .

Heaven prosper it! may peace, and love,
And hope, and consolation, fall,
Through its meek influence, from above,
And penetrate the hearts of all;
All who, around the hallowed Fane,
Shall sojourn in this fair domain:
Grateful to Thee, while service pure,
And ancient ordinance, shall endure,
For opportunity bestowed
To kneel together, and adore their God!

RYDAL MOUNT, the residence of William Wordsworth, Esq., for the last thirty-seven years of his life, stands a little to the north-east of the Church:—

* Bekangs Ghyll—or the dell of Nightshade—in which stands St. Mary's Abbey, in Low Furness.

"Low and white, yet scarcely seen
Are its walls for mantling green;
Not a window lets in light
But through flowers clustering bright;
Not a glance may wander there,
But it falls on something fair;
Garden choice and fairy mound,
Only that no elves are found;
Winding walk and sheltered nook,
For student grave, and graver book;
Or a bird-like bower, perchance,
Fit for maiden or romance."

Miss JEWSBURY.

This little paradise has so long been associated with the name of the Poet Laureate of England, that the following account of it, extracted from his "Memoirs," recently published, will, we doubt not, be read with interest.

"The house stands upon the sloping side of a rocky hill, called Nab Scar. It has a southern aspect: in front of it is a small semicircular area of grey gravel, fringed with shrubs and flowers, the house forming the diameter of the circle. From this area there is a descent by a few stone steps southward, and then a little ascent to a grassy mound. Here let us rest a little. At our back is the house; in front, rather to the left in the horizon, is Wansfell, to which the Poet has paid a grateful tribute in two of his later Sonnets (42 and 43).

"Wansfell! this household has a favoured lot,
Living with liberty on thee to gaze."

Beneath it, the blue smoke shows the place of the town of Ambleside. In front, is the Lake of Windermere, shining in the sun; also, in front, but more to the right, are the fells of Loughrigg, one of which throws up a massive solitary crag, on which the Poet's imagination pleased itself to plant an imperial castle:

"Aerial rock whose solitary brow,
From this low threshold daily meets the sight."

Looking to the right in the garden, is a beautiful glade, overhung with rhododendrons in most luxuriant leaf and bloom. Near them is a tall ash tree, in which a thrush has sung for hours together for many years. Not far from it is a laburnum, in which the osier cage of the doves was hung. Below, to the west, is the vegetable garden, not parted off from the rest, but blended with it by parterres of flowers and shrubs.

"Returning to the platform of grey gravel before the house, we pass under the shade of a fine sycamore, and ascend to the westward by fourteen steps of stones, about nine feet long, in the insterstices of which grow the yellow flowering poppy and the wild geranium or Poor Robin,

> "Gay,
> With his red stalks upon a sunny day,"

a favourite with the Poet, as his verses show. The steps above mentioned lead to an upward *sloping* TERRACE, about two hundred and fifty feet long. On the right side it is shaded by laburnums, Portugal laurels, mountain ash, and fine walnut trees and cherries; on the left it is flanked by a low stone wall, coped with rude slates, and covered with lichens, mosses, and wild flowers. The fern waves on the walls, and at its base grows the wild strawberry and foxglove. Beneath this wall, and parallel to it, on the left, is a *level* TERRACE, constructed by the Poet for the sake of a friend most dear to him and his, who, for the last twenty years of Mr. Wordsworth's life, was often a visitor and inmate of Rydal Mount. This terrace was a favourite resort of the Poet, being more easy for pacing to and fro, when old age began to make him feel the acclivity of the other terrace to be toilsome. Both these terraces command beautiful views of the vale of Rothay and the banks of the Lake of Windermere.

"The *ascending* TERRACE leads to an arbour lined with fir cones, from which, passing onward, on opening the latched door, we have a view of the lower end of Rydal Lake, and of the long, wooded, and rocky hill of Loughrigg, beyond and above it. Close to this arbour-door, is a beautiful sycamore, with five fine Scotch firs in the fore-ground, and a deep bay of wood to the left and front, of oak, ash, holly, hazel, fir, and birch. The terrace-path here winds gently off to the right, and becomes what was called by the Poet and his household the FAR TERRACE on the mountain's side:

> "The Poet's hand first shaped it, and the steps
> Of that same bard—repeated to and fro,
> At morn, at noon, and under moonlight skies,
> Through the vicissitudes of many a year—
> Forbad the weeds to creep o'er its grey line."

Here he

> "Scattered to the heedless winds
> The vocal raptures of fresh poesy;

And here he was often

"locked
In earnest converse with beloved friends."

"The 'far terrace,' after winding along in a serpentine line for about one hundred and fifty feet, ends at a little gate, beyond which is a beautiful well of clear water, called 'the Nab well,' which was to the Poet of Rydal—a professed water-drinker—what the Bandusian fount was to the Sabine bard:

"Thou hast cheered a simple board
With beverage pure as ever fixed the choice
Of hermit dubious where to scoop his cell,
Which Persian kings might envy."

"Returning to the arbour we descend by a narrow flight of stone steps to the kitchen-garden, and passing through it southward, we open a gate and enter a field sloping down to the valley, called, from its owner's name, 'Dora's field.' Not far on the right, on entering this field, is the stone bearing this inscription:

"In these fair vales hath many a tree
At Wordsworth's suit been spared;
And from the builder's hand this stone,
For some rude beauty of its own,
Was rescued by the Bard."

And the concluding lines will now be read with pathetic interest:

"So let it rest; and time will come,
When here the tender-hearted
May heave a gentle sigh for him
As one of the departed."

"Near the same gate, we see a pollard oak, on the top of whose trunk may yet be discerned some leaves of the primrose which sheltered the wren's nest:

——"She who planned the mossy lodge,
Mistrusting her evasive skill,
Had to a primrose looked for aid,
Her wishes to fulfil."

"On the left of this gate, we see another oak, and beneath it a pool, to which the gold and silver fish, once swimming in a vase in the library of the house, were transported for the enjoyment of greater freedom:—

"Removed in kindness from their glassy cell
To the fresh waters of a living well;
An elfin pool, so sheltered that its rest
No winds disturb."

Passing the pool, and then turning to the right, we come to some stone steps leading down the slope; and to the right, engraven on the rock, is the following inscription, allusive to the character of the descent:—

> "Wouldst thou be gathered to Christ's chosen flock,
> Shun the broad way too easily explored,
> And let thy path be hewn out of the Rock
> The living Rock of God's eternal Word."

"The house itself is a modest mansion, of a sober hue, tinged with weather stains, with two tiers of five windows; on the right of thèse is a porch, and above, and to the right, are two other windows; the highest looks out of what was the Poet's bed-room. The gable-end at the east—that first seen on entering the grounds from the road—presents on the ground-floor the window of the old hall or dining-room. The house is mantled over here and there with roses and ivy, and jessamine and Virginia creepers.

"In this cottage Wordsworth died on the same day of the month as that on which Shakspeare was born, April 23rd, being also the day of Shakspeare's death. On Saturday, the the 27th, 1850, his mortal remains, followed to the grave by his own family and a very large concourse of persons of all ranks and ages, were laid in peace, near those of his children, in Grasmere church-yard. His own prophecy, in the lines

> "Sweet flower! belike one day to have
> A place upon thy Poet's grave,
> I welcome thee once more,"

is now fulfilled. He desired no splendid tomb in a public mausoleum. He reposes, according to his own wish, beneath the green turf, among the dalesmen of Grasmere, under the sycamores and yews (probably planted by his own hand*) of a country church-yard, by the side of a beautiful stream, amid the mountains which he loved; and a solemn voice seems to breath from his grave, which blends its tones in sweet and holy harmony with the accents of his poetry, speaking the accents of humility and love, of adoration and faith, and preparing the soul, by a religious exercise of the kindly affections, and by a devout contemplation

* Vide "Memoirs," p. 41, Vol. I., and p. 266 Vol. II.

of natural beauty, for translation to a purer, and nobler, and more glorious state of existence, and for a fruition of heavenly felicity."

A plain blue head-stone marks the grave of the Poet, without any inscription but his name; and in the Church is a neat marble monument to his memory, bearing the following epitaph:—

To the Memory of
WILLIAM WORDSWORTH,
a true Philosopher and Poet,
who, by the special gift and calling of
Almighty God;
whether he discoursed on Man or Nature,
failed not to lift up the heart
to holy things,
tired not of maintaining the cause
of the poor and simple;
and so, in perilous times was raised up
to be a chief minister
not only of noblest poesy,
but of high and sacred truth,
THIS MEMORIAL
is placed here by his Friends and Neighbours,
in testimony of
Respect, Affection, and Gratitude.
Anno 1851.

RYDAL WATER

Is one of the smallest of the English Lakes, but certainly one of the most beautiful, from its woody islets and picturesque shores; but it ought to be observed here, that Rydal-mere is no where seen to advantage from the *main road.* Fine views of it may be had from Rydal Park; but these grounds, as well as those of Rydal Mount (Mrs. Wordsworth) and Ivy Cottage, now called Glen Rothay (Wm. Ball, Esq.) from which also it is viewed to advantage, are private. A foot-road passing behind Rydal Mount and Nab Scar to Grasmere, is very favourable to views of the lake and the vale, looking back towards Ambleside. The horse-road, also, along the western side of the lake, under Loughrigg fell, as before mentioned, does justice to the beauties of this small mere, of which the traveller who keeps the high road is not at all aware.

About 200 yards beyond the last house on the Keswick side of Rydal village, the road is cut through a low wooded rock, called Thrang Crag. The top of it, which is only a few steps on the south side, affords the best view of the vale which is to be had by a traveller who confines himself to the public road.

A short distance beyond this crag, proceeding towards Grasmere, a neat cottage by the road side will attract the notice of the tourist. That cottage is called THE NAB, to which a certain degree of interest is attached as being for some years the residence of HARTLEY COLERIDGE.* Here he died on Saturday, the 6th of January, 1849, and was interred on the following Thursday, in the south-east angle of Grasmere Church-yard, the entrance to which from the north is by a lych-gate, under which you pass to the village school. "Possibly," says his biographer, "this thought may have been in my brother's mind, and an image of this quiet resting-place in his mind's eye, when he penned the following characteristic observations on the choice of a grave, which were found written on the margin of an old number of the London Magazine:

"I have no particular choice of a church-yard, but I would repose, if possible, where there are no proud monuments, no new-fangled obelisks or mausoleums, heathen in everything but taste, and not Christian in that. Nothing that beto-

---

* "After the cheerfulness of the Mount, the residence of Wm. Wordsworth, which lies high above it, at the distance of a few furlongs," says a recent writer, "this cottage looks lone and desolate." To this observation Coleridge's biographer replies, 'Lonely it is; but not, to my feelings, 'desolate.' It stands by itself on the road-side, between Ambleside and Grasmere, and at nearly an equal distance between the two, having the little lake of Rydal, with its two woody islets in front, at the distance of a stone's throw from the door. A sloping meadow behind leads to the many-coloured side of Nab Scar, which rises steep and, in part, precipitous, through a skirt of trees, with which it is slightly feathered to a considerable height. On the opposite side of the lake runs the range of Loughrigg, greeting the eye with a rich variety of hue and outline, light and shade. On the whole, I take the character of the place to be that of cheerful retirement without seclusion, well fitted for the abiding place of a man at once contemplative and social, who living much alone, and in communion with Nature, yet needed ready access to the haunts of men—and such was my brother.—It was surely a happy, and, so to speak, a suitable disposition of events—I would not lightly use the word Providential—which brought my brother to spend his latter days, as it were, under the shadow and at the foot of that great poet, his father's friend,—so pronounced in words of immortal fame,—with whom his own infancy and boyhood had been so closely and so affectionately linked. As a poet, he would have accounted this an honourable place, and would have claimed no higher. To this, of all his contemporaries, he was every way best entitled. Living in such neighbourhood together, and with no greater distance of affection, they were not far divided in their deaths, and now they lie all but side by side."

keneth aristocracy, unless it were the venerable memorial of some old family long extinct. If the village school adjoined the church-yard, so much the better. But all this must be as He will. I am greatly pleased with the fancy of Anaxagoras, whose sole request of the people of Lampsacus was, that the children might have a holiday on the anniversary of his death. But I would have the holiday on the day of my funeral. I would connect the happiness of childhood with the peace of the dead, not with the struggles of the dying."

A neat monument of Caen stone marks the grave of Hartley Coleridge, with the following short epitaph:—

"By thy Cross and Passion, good Lord deliver us."
HARTLEY COLERIDGE,
born September 19th, A. D. 1796, deceased January 6th, A. D. 1849.

And on a stone at the foot of the grave is the following inscription:—

The Stones which mark the Grave of HARTLEY COLERIDGE, eldest son of SAMUEL TAYLOR COLERIDGE, were erected by his surviving Brother and Sister, towards the close of the year 1850.

From Nab Cottage to White Moss Slate Quarries it is barely a mile, and here the pedestrian should take the old road over the hill, for the sake of the fine retrospective views of Rydal which it affords, and for the more favourable view of Grasmere, which he is now about to approach. On this road he will pass a gate on his left, which, time out of mind, has been called the WISHING GATE, from a belief that wishes formed or indulged there have a favourable issue. This road will also conduct him through that part of the village called TOWN END, passing on his right the cottage in which Wordsworth took up his abode on his first settlement at Grasmere in the year 1799, and which still retains the form it then wore. The front of it faces the lake; behind is a small plot of orchard and garden ground, the enclosure shelving upward toward the woody sides of the mountains above it. Many of his Poems, as the reader will remember, are associated with this fair spot:—

"This plot of orchard ground is ours,
My trees they are, my sister's flowers."

His feelings are thus expressed in settling in his new house, and in looking down from the hills which embosom the lake:—

"On Nature's invitation do I come,
By Reason sanctioned. Can the choice mislead,
That made the calmest, fairest spot on earth,
With all its unappropriated good,
My own? . . . . . .

. . . .
Embrace me then, ye hills, and close me in,
Now in the clear and open day I feel
Your guardianship! I take it to my heart;
'Tis like the solemn shelter of the night.
But I would call thee beautiful; for mild,
And soft, aad gay, and beautiful thou art,
Dear valley, having in thy face a smile,
Though peaceful, full of gladness. Thou art pleased,
Pleased with thy crags, and woody steeps, thy lake,
Its own green island, and its winding shores,
The multitude of little rocky hills,
Thy Church, and cottages of mountain stone
Clustered like stars some few, but single most,
And lurking dimly in their shy retreats,
Or, glancing at each other cheerful looks,
Like separated stars with clouds between."

The new road skirts the margin of the lake, but is fenced from it by an odious stone wall, and joins the old road at Town-end before mentioned, from whence the road into the village takes the left hand.

GRASMERE

Is beautifully situated at the northern end of the lake, which is more than a mile in circumference, and contains one bare island. The Church, an ancient structure dedicated to St. Oswald, will claim the notice of the tourist, from being that to which the following beautiful lines by Mr. Wordsworth in his Poem of "The Excursion," were intended by him to apply.*

"Not raised in nice proportions was the pile,
But large and massy, for duration built;
With pillars crowded, and the roof upheld
By naked rafters, intricately crossed
Like leafless underboughs, 'mid some thick grove,
All wither'd by the depth of shade above.

* "The Church noticed in the Excursion is that of Grasmere. The interior of it has been improved lately—made warmer by undrawing the roof, and raising the floor; but the rude and antique majesty of its former appearance has been impaired by painting the rafters; and the oak benches, with a simple rail at the back dividing them from each other, have given way to seats that have more the appearance of pews. It is remarkable that, excepting only the pew belonging to Rydal Hall, that to Rydal Mount, the one to the Parsonage, and, I believe, another, the men and women still continue, as used to be the custom in Wales, to sit on separate sides of the Church, from each other. Is this practice as old as the Reformation? and when and how did it originate?"—See "MEMOIRS," p. 39. Vol. II.

0 1 2 3 4 5 6 7 8

Scale of Miles.

Published by J. Hudson, Kendal.

Admonitory texts inscribed the walls—
Each in its ornamental scroll enclosed,
Each also crown'd with winged heads—a pair
Of rudely-painted cherubim. The floor
Of nave and aisle, in unpretending guise,
Was occupied by oaken benches, ranged
In seemly rows."

There are two small inns in the vale of Grasmere, one near the Church (*The Red Lion*), the other (*The Swan*) on the main road. From the former the valley may be more conveniently explored in every direction, and a mountain walk taken up Easedale to Easedale Tarn (2½ miles), one of the finest tarns in the country, thence to Stickle Tarn and to the top of Langdale Pikes. See also the vale from Butterlip How, half a mile from the inn. It is the finest elevation of moderate height in the neighbourhood. Helm Crag may be visited from Grasmere. It is two miles to its summit, which is extremely rugged, and the ascent is somewhat difficult. The shattered apex of this mountain, as seen from certain points in the valley, bears a striking resemblance to a lion couchant, with a lamb lying at the end of his nose; and to an old woman cowering.* ALLAN BANK, the residence of the Rev. — Jeffries, is only a short distance out of the road leading from *The Red Lion* to Easedale, and from some places in the avenue Helm Crag is a pleasing object. Seat Sandal and all the lofty mountains south of it are are seen towering over the pretty undulating Butterlip How and other elevations,

* Mr. Wordsworth, in one of his Poems on the Naming of Places, entitled "Johanna," thus introduces the old lady :—

When I had gazed perhaps two minutes space,
Johanna, looking in my eyes, beheld
That ravishment of mine, and laughed aloud.
The Rock, like something starting from a sleep,
Took up the Lady's voice, and laughed again;
That ancient woman seated on Helm-crag
Was ready with her cavern; Hammar-scar
And the tall steep of Silver-how, sent forth
A noise of laughter: southern Loughrigg heard,
And Fairfield answered with a mountain tone;
Helvellyn far into the clear blue sky
Carried the Lady's voice; old Skiddaw blew
His speaking-trumpet;—back out of the clouds
Of Glaramara southward came the voice;
And Kirkstone tossed it from its misty head.

and the whole Vale of Grasmere is hardly anywhere seen to greater advantage than from this point.

A boat is kept by the innkeeper, and this circular vale, in the solemnity of a fine evening, will make, from the bosom of the lake, an impression that will scarcely ever be effaced.

The steep and rugged bridle-road from Grasmere to Patterdale, by Grisedale Tarn, a distance of seven miles, turns off at a smithy four miles and three quarters from Ambleside.

Beyond the toll-bar the road begins to ascend the Pass of Dunmail Raise, between Steel Fell on the west, and Seat Sandal on the east. At the highest point, which is 720 feet above the sea, it passes a low cairn, or pile of stones, said to have been raised in the year 945, by the Anglo-Saxon King Edmund, after the defeat and death, on this spot, of Dunmail (or Dumhnail) the British King of Cumbria, and the consequent destruction of that kingdom. The river on the right of the Raise divides the counties, whence to the Nag's Head, Wythburn, is one mile and a quarter. Opposite the inn is a small Chapel, "as lowly as the lowliest dwelling." HARTLEY COLERIDGE has the following verses on this quiet spot:—

Here, traveller, pause and think, and duly think,
What happy, holy thoughts may heavenward rise,
Whilst thou and thy good steed together drink,
Beneath this little portion of the skies.

See! on one side, a humble house of prayer,
Where Silence dwells, a maid immaculate,
Save when the Sabbath and the priest are there,
And some few hungry souls for manna wait.

Humble it is and meek and very low,
And speaks its purpose by a single bell;
But God Himself, and He alone, can know
If spiry temples please Him half so well.

Then see the world, the world in its best guise,
Inviting thee its bounties to partake;
Dear is the Sign's old time-discolour'd dyes,
To weary trudger by the long black lake.

And pity 'tis that other studded door,
That looks so rusty right across the way,
Stands not always as was the use of yore,
That whoso passes may step in and pray.

This is a convenient Station for asceding Helvellyn, and the mountain track approaching it may be observed from the door of the inn. The stream that the tourist will be directed to follow issues from a small well, called Brownrigg's well, only a few hundred yards to the south of the summit, and is therefore perhaps the best guide that he can have, unless he takes a professional one from the inn. Another favourable point for commencing the ascent of this mountain is at the sixth milestone from Keswick. The ascent of Helvellyn will be hereafter noticed in the Patterdale Excursions.

The direct road from Grasmere to Keswick does not (as has been observed of Rydal-mere) show to advantage

THIRLMERE,

or Wythburn Lake, with its surrounding mountains. By a traveller proceeding at leisure, a deviation ought to be made from the main road, when he has advanced a little beyond the sixth milestone short of Keswick, from which point there is a noble view of the Vale of Legberthwaite, with Blencathra (commonly called Saddleback) in front. Having previously enquired, at the inn near Wythburn Chapel, the best way from this milestone to the bridge that divides the lake, he must cross it, and proceed with the lake on the right, to the hamlet a little beyond its termination, and rejoin the main road upon Shoulthwaite Moss, about four miles from Keswick; or, if on foot, the tourist may follow the stream that issues from Thirlmere down the romantic Vale of St. John's, and so (enquiring the way at some cottage) to Keswick, by a circuit of little more than a mile. By following the direct road, and when about a mile from Keswick, at the top of Castlerigg Brow, one of the richest mountain scenes is gradually unfolded that can be enjoyed from any of the carriage roads in the North of England. A more interesting tract of country is scarcely anywhere to be seen, than the road between Ambleside and Keswick, with the deviations that have been pointed out.

From AMBLESIDE, through Grasmere, Easedale, Greenup, and Borrowdale to KESWICK.

4	Grasmere Church	4	1	Greenup Dale Head	10
1	Goody Bridge	5	3	Down Greenup vale to Stonethwaite...	13
1	Thorneyhow	6	7	Keswick...	20
1	Far Easdale	7			
2	Wythburn Dale Head... ...	9			

Pursue the road, as before described, as far as Grasmere, from whence "the valley of Easedale runs far into the northern hills on the western side of Helm Crag. Near its mouth a stream flows from Easedale Tarn, and from the whiteness of the broken water is called Sour-milk Gill. Up this seldom-visited glen the foot traveller may pursue his way from Grasmere to Keswick, ascending by a steep and laborious climb to a narrow level tract of moor called Colddale Fell; after which he will descend into the Stonethwaite branch of Borrowdale, nor will he regret, though the way be longer and far more laborious, having exchanged the high road for the freedom of the mountain-side."

From AMBLESIDE, through Great Langdale, to the STAKE, and thence through Borrowdale, to KESWICK.

5	Langdale Chapel	5	5	Stonethwaite	17
2	Lisle Bridge, near Dungeon Gill	7	1	Rosthwaite	18
1	Langdale Head	8	1	Bowder Stone	19
4	Top of the Stake	12	5	Keswick	24

The finest approach to Great Langdale is by pursuing the Keswick road to Pelter Bridge (one mile), which having crossed, pass on the side of the Rothay by Coat How to Rydal and Grasmere lakes, thence by High Close and Langdale Chapel to Lisle Bridge and Millbeck, which places have been before noticed in the Langdale Excursion. Ascending the Stake, the road is on the side of a turbulent stream, which dashes down into the valley of Langdale. Half a mile beyond the top of the Langdale Stake, begins the descent into Borrowdale by the side of a river through the valley of Langstreth, where all is in a state of wildness and desolation. At the top of the Stake is a grand exhibition of the high summits of Bow Fell, Hanging Knotts, Scawfell Pikes, and Great Gable, and at a considerable distance is seen Skiddaw, partly intercepted by nearer mountains. Half way down the vale the road crosses the river, having, in the direction of Stonethwaite, a large and curious stone on the right, called Black Cap, above which is Sergeant Crag, and nearer Stonethwaite is the bold rocky elevation of Eagle Crag on the right. From Stonethwaite, the road to Keswick is by Rosthwaite, in Borrowdale, where there is a small public-house. Thence pass Bowder Stone, Lowdore, and Barrow, which will hereafter be described in the Keswick Excursions.

From AMBLESIDE, over Wrynose and Hardknott, to WAST WATER, thence by Sty Head to KESWICK, or return to Ambleside by Sty Head Tarn through Langdale, or by Seathwaite, through Eskdale.

Stage	Total
1 Clappersgate	1
2 Skelwith Bridge	3
1 Colwith Bridge and Force ...	4
2 Fell Foot	6
2 Top of Wrynose	8
2 Cockley Beck	10
2 Hardknott Castle	12
4 Dalegarth Hall and Stanley Gill	16
1 Road on the left by Ulpha to Broughton	17
3 Santon Bridge	20
2 Strands Public House	22
[From Santon Bridge direct to Crook, at the foot of Wast Water, is 1 mile.]	
3 Netherbeck Bridge	25
1 Overbeck Bridge	26
1 Wastdale Head	27
2 Sty Head	29
12 Keswick by Bowder Stone ...	41
From Sty Head to Ambleside by Sty Head Tarn, Sprinkling Tarn, and Angle Tarn, and thence through the Vale of Langdale, 16½ miles, making the round	45½
From Sty Head by Seathwite, and thence through Greenup and Eskdale, to Ambleside, 18 miles, making the round	47

This road is by Skelwith and Colwith Bridges, at which latter place there is a fine Waterfall, called Colwith Force, scarcely inferior to any in the country, and thence through Little Langdale. It has been described in the Langdale Excursion as far as the place where it diverges to Blea Tarn and Great Langdale, a distance of scarcely seven miles from Ambleside. (p. 40.) Hence the road is to Fell Foot, formerly a public-house, when this was the main road from Kendal to Whitehaven, a fact which those who now travel over it will find it hard to believe. At the time we are speaking of, the only mode for the conveyance of goods was on the backs of pack-horses, long trains of which were often to be seen traversing these hills.* At Fell Foot begins the ascent of Wrynose to the three *Shire Stones,* where the counties of Cumberland, Westmorland, and Lancaster unite on the top of the hill. Here the road enters Lancashire, having the stream which divides it from Cumberland, on the right, and descends, though not abruptly, upon Cockley Beck, only to cross the valley and climb another mountain no less high and difficult of ascent, called Hardknott, which separates Seathwaite from Eskdale. "The ascent of the upper part of the valley at Cockley Beck, where it is crossed by the mountain-road of which we have been speaking, is dreary. A tract of desolate hills, nurses of the Esk and Duddon, rises towards the north-west into the lofty range of Scawfell and Bowfell. The

* Bells were attached to the collar of the *leading* horse of the train. A collar of this kind may be seen in the Museum at Kendal.

head of Eskdale lies between these, the highest and the roughest mountains in the country; and we might here fancy ourselves deep in the recesses even of the wilder parts of the Scottish Highlands. The precipices of Scawfell, and of the higher point of that great mountain, called The Pikes, tower darkly and awfully on the western side; and even on the eastern, where Bowfell slopes down more gently, the passage of the traveller must be slow and cautious. The assistance of an experienced guide in this wild and perplexing region is strongly recommended. No precipice, however, bars up the head of the dale, which rises gradually to the green ridge that marks the water's source between Eskdale and Borrowdale. This height, itself a depression between Great End and that part of Bowfell, called Hanging Knott, is called Esk Hause.* From it we look directly down the whole of Borrowdale, and command a view of Derwentwater, with its specks of islands, the whole closed by the pyramidal group of Skiddaw, which is here seen from head to foot, and to the greatest advantage The outbreak of the river from this upland glen to the lower valley, some five or six miles from Esk Hause, forms a succession of falls and rapids for a considerable distance, fringed with birch and mountain ash, the first signs of better soil and milder climate. These, in their varied combinations of rock and water, furnish ample studies for the artist or sketcher."

Something more than half way down the hill in descending into Eskdale, about 120 yards on the right of the road, are the remains of *Hardknott Castle*, mentioned in p. 15, from whence there is a magnificent view of Scawfell and the Pikes, supported by the immense buttresses rising from the Esk. At the foot of the hill there is a very extensive sheep-farm on the right called Brotherilkeld, and one on the left called Toes. "Proceeding down the valley, BIRKER FORCE is seen dashing over the rocks on the left, and about two miles from the foot of the hill we come to a public-house at Bout; within a mile of which is situated a very fine waterfall called STANLEY GILL, far up a deep, narrow, and thickly-wooded ravine. The stream is small, and in height the fall is not remarkable; but in the picturesque character of its accompaniments it is inferior to none of those that are better

* Pronounced *Ash Course* by the dalesmen.

known in the country." The road to this waterfall turns off on the left at the village school, and a guide to the fall may be had at Dalegarth Hall, a farm-house close at hand [*See Note* p. 15] From the hamlet of Bout the main road should be followed nearly to Santon Bridge, where it turns off to the right to the Strands at Nether Wastdale, a distance of two miles and a half, where there are two small inns. There is a nearer cut to the Strands for pedestrians, by a foot-road through Mitredale, which strikes across the hill on the right, a little before reaching Santon Bridge.

From Bout there is a rough mountain-road which traverses the moor to Wastdale Head, passing a cheerless sheet of water called Burnmoor Tarn, between Scawfell and the Screes, and then descends down a steep peat-track into Upper Wastdale, a little above the lake. From Westdale Head the road is on the western side of Wastwater to the Strands. The eastern side of the lake is skirted by the Screes, and is not only difficult but dangerous to attempt, from the loose and crumbling nature of the materials of which it is composed. Tourists tarrying here for a day or two will find many pleasant excursions in the neighbourhood. From a little hill called Latterbarrow is a good general view of the surrounding country: but from a hill by the Gosforth road, near the inn, is the best general view of the mountains. From Latterbarrow the lake can be seen, but Scawfell and the Pikes are shut out from this point of view.—[*See the Plate I. of Sketches of the Mountains.*]

Calder Abbey, a small but beautiful ruin, is eight miles from the Strands, but this place is more generally visited in going from Wastdale by Ennerdale Water, Lowes Water, and Scale Hill to Keswick.

"There is a simplicity and severity about Wast Water not to be found in any of its neighbour lakes, except perhaps that of Ennerdale, which is equally destitute of the cheerfulness imparted by cultivation, but inferior in the height and ruggedness of its mountain boundaries." It is three miles long, half a mile broad, and forty-five fathoms in depth, being deeper than any of the other lakes. "Within some half an hour's walks from Strands is a remarkable spot called Haul-gill, or else Hollow-gill. It is a deep ravine at the south-west foot of the Screes, among granite rocks, which, by the decomposition of their felspar, have been wasted into abrupt peaks and precipices—a sort of miniature mimicry of the *aiguilles* of Chamouni. This is one of the most curious and striking things in the whole district; it is a good place for ascending the Screes from Nether Wastdale (as the valley below the lake is called) for those who have strong nerves. There is a very beautiful vein of spicular iron ore here; also some fine hæmatite."

On the way from the Strands by Gale and Crookhead Cottages, the residences of the Messrs. Rawson, which the tourist must now pursue on his road to Keswick by Sty Head, the Screes are occasionally in view, from whence the Great Gable is seen in the vista formed by Middle Fell, Yewbarrow, and Kirkfell on the left, and on the right by Lingmell and the north end of the Screes. As you advance toward the head of the lake the pastoral valley of Bowderdale is on the left, stretching up towards the Haycocks. From hence Scawfell is a commanding object, and the Pikes begin to shew their separation by the gradual development of the deep chasm called Mickle Door, which divides their summits.

From Wastdale Head, a sequestered hamlet, with a chapel,* but no inn,† you commence a precipitous ascent to Sty Head, the highest Pass in the district, having the huge rocks of Great Gable on the left, and those of Lingmell Crag on the right: in front, Great End. Lingmell Crag is succeeded by Broad Crag, and the Pikes tower majestically over the whole. From Sty Head the road descends by a horse track through Seathwaite and Borrowdale to Keswick, a distance of twelve miles. At Seathwaite, three miles and a half from Sty Head, the tourist, should he require refreshment, will meet with good and homely fare at Mrs. Dixon's hospitable board. The objects on this road will be more particularly noticed hereafter, in the walk to Sty Head from Keswick. From Sty Head the road to Ambleside is either by leaving Sprinkling Tarn on the left and Angle

* This chapel is perhaps the humblest specimen of ecclesiastical architecture in the kingdom. The edifice contains eight pews, and is lighted by three small windows—one at the eastern end and one on each side of the building. Beside these, there is a small sky-light immediately over the pulpit!

† Tourists can be accommodated with plain and wholesome refreshments at Ritson's, a clean and comfortable farm-house.

Tarn on the right hand, and proceeding through Langdale; or, through Borrowdale and Stonethwaite, thence over Greenup through Eskdale and Grasmere.

From the hamlet of Wastdale Head there is a rough foot-road through the valley of Mosedale, which stretches westward between the mountains of Kirkfell and Yewbarrow, into Gillerthwaite at the head of Ennerdale dale, and thence by the Pass of Scarf-gap to Gatesgarth at the head of Buttermere. Having gained the head of Mosedale the road crosses a hollow on the right between Kirkfell and the Pillar, and descends rapidly with the stream on the right into Gillerthwaite, which is closed in at the head by Kirkfell and Great Gable. On the opposite side of the valley, High Stile and Red Pike seperate it from Buttermere. The small stream called the Liza, is crossed at the sheep-fold, and must be followed downwards for a short space, where an indistinct path over a second hollow, between the Haystacks and High Crag, called Scarf Gap, must be pursued, which brings the traveller to Gatesgarth, at the head of Buttermere. This route from the Strands to Buttermere, comprises a great variety of scenery, and is perhaps one of the finest mountain walks in the district. As the path is ill-marked in many places, it would be prudent to take a guide. In the Autumn of 1842 an inexperienced tourist undertook this route, and started from Wastdale Head without a guide. After wandering about for some time he missed the road, and, instead of getting into Buttermere by the Pass of Scarf Gap, he took the deep ravine between Kirkfell and the Gable, and arrived (without finding out his mistake) at the precise point from which he had started several hours before having made a circuit of many miles!

It may be observed that the ascent of Scawfell may be made with less exertion and fatigue from the Strands than from any other Station. A boat may be taken to the head of the lake, where the ascent commences at once upon Lingmell, and, with a guide to point out the way, the distance to the summit is about three miles, and may be accomplished in an hour and a half by active pedestrians. A remarkable gill, called PEASGILL, situate on the nort-west side of Lingmell, might be visited in the descent. The ascent of Scawfell is, however, more frequently made from Borrowdale than from any other point, and will, therefore, be fully noticed in the Keswick Excursions.

KESWICK.

KESWICK is a small market-town delightfully situated near the foot of Derwentwater. Tourists generally make Keswick their head-quarters for a time, and are there provided with good accommodations and the requisities for their excursions. INNS, *Royal Oak and Queen's Head* The principal manufactures of Keswick consist of black-lead pencils, coarse woollens, flannels, &c. The mineral black-lead (*Plumbago*) of which pencils are manufactured, is found in the mines of Borrowdale, and although these mines are in the vicinity of Keswick, the pencil-makers are obliged to purchase all their material at the Company's wharehouse in London, whither it is sent in casks, and exposed for sale only on the first Monday in every month. There are in Keswick two Museums, illustrating, in addition to many foreign curiosities, the natural history and mineral productions of the surrounding country. At each of these the visitor can purchase geological specimens from the rocks of the neighbourhood. An accurate Model of the Lake District, ingeniously constructed by Mr. Flintoft, is also exhibited here in the summer season, and is well worth a careful examination. The horizontal and vertical scale of this Model is three inches to a mile; in length, from Sebergham to Rampside, 51 miles, or 12 feet 9 inches; breadth, from Shap to Egremont, 37 miles, or 9 feet 3 inches; circumference, exclusive of sea, 176 miles. The coast is shewn two-fifths of the distance, presenting the Bays of Morecambe, Duddon, and Ravenglass. The inspector has before him the whole chain of mountains in the Lake District, in three principal groups—the Scawfell, the Helvellyn, and the Skiddaw group, with their numerous interesting valleys, spotted with sixteen large lakes. On the uplands are seen fifty-two small ones, principally high in the mountain recesses, surrounded by contorted and precipitous rocks. On this Model are marked the towns of Kendal, Ambleside, Ulverston, Bootle, Broughton, Cockermouth, Keswick, Penrith, and Shap. The face of the whole is coloured to nature, with the exception of the churches, which are coloured red. The plantations are raised, and coloured dark green; the rivers, lakes, and sea, light blue; roads, light brown; and the houses white, as they usually appear. A new Church was recently built at the south end of the town by the late John Marshall, Esq.,

Published by J Hudson, Kendal

the purchaser of the estates in this vale which belonged to Greenwich Hospital. A Parsonage and School-house have, since his decease, been added by the family of Mr. Marshall, of Hallsteads. The church is an elegant structure, delightfully situated on a gentle eminence, from which an extensive panoramic view of the surrounding country may be had. The Parish Chuch, called Crosthwaite Church, is a mile from the town, in the opposite direction. It is an ancient edifice, consisting of a Nave, two lateral aisles and a porch. The interior was completely remodelled and highly embellished a few years ago at a considerable cost, a great portion of which was borne by James Stanger, Esq., a neighbouring resident gentleman. In this Church there is a handsome Monument in white marble, by Lough, to the memory of SOUTHEY, which consists of a recumbent figure of the Poet at full length raised on a pedestal of Caen stone, and as a faithful likeness, and a work of art has great merit. It is said to have cost £1100, which was raised by public subscription. The following inscription on the monument is by Southey's old and valued friend Wordsworth:—

> Ye Vales and Hills, whose beauty hither drew
> The Poet's steps, and fixed them here—on you
> His eyes have closed! and ye loved books, no more
> Shall SOUTHEY feed upon your precious lore,
> To works that ne'er shall forfeit their renown
> Adding immortal labours of his own—
> Whether he traced Historic Truth, with zeal
> For the State's guidance, or the Church's weal;
> Or Fancy disciplined with studious Art
> Informed his pen, or wisdom of the heart,
> Or judgment sanctioned in the Patriot's mind,
> By reverence for the rights of all mankind.
> Wide were his aims, yet in no human breast,
> Could private feelings find a holier rest,
> His joys, his griefs, have vanished like a cloud
> From Skiddaw's top; but he to Heaven was vowed
> Through a life long and pure; and Christian Faith
> Calmed in his soul the fear of Change and Death.

The grave of Southey is in the Church-yard, to which the stranger will be conducted by a well-trodden path.

GRETA HALL, the residence of Southey, for the last forty years of his life, will possess some interest to the literary tourist. It stands, at the northern extremity of the town, a few hundred yards only to the right of the Bridge. It commands a fine view of the scenery of the valley, which the Poet himself has sketched in the following beautiful lines:—

'Twas at that sober hour, when the light of day is receding,
And from surrounding things the hues wherewith day has adorned them
Fade, like the hopes of youth, till the beauty of earth is departed:
Pensive, though not in thought, I stood at the window, beholding
Mountain and lake and vale; the valley disrobed of its verdure;
Derwent retaining yet from eve a glassy reflection,
Where his expanded breast, then still and smooth as a mirror,
Under the woods reposed: the hills that, calm and majestic,
Lifted their heads in the silent sky, from far Glaramara,
Bleacrag and Maidenmawr, to Grizedale and westernmost Wythop;
Dark and distinct they rose. The clouds had gather'd above them
High in the middle air, huge purple pillowy masses,
While in the west beyond was the last pale tint of the twilight;
Green as a stream in the glen, whose pure and chrysolite waters
Flow o'er a schistous bed; and serene as the age of the righteous.
Earth was hushed and still; all motion and sounds were suspended;
Neither man was heard, bird, beast, nor humming of insect,
Only the voice of the Greta, heard only when all is in stillness.
Pensive I stood and alone, the hour and the scene had subdued me,
And as I gazed in the west, where Infinity seem'd to be open,
Yearn'd to be free from time, and felt that this life is a thraldom.

DERWENT WATER

Is upwards of three miles in length, and a mile and a half in its greatest breadth. It is adorned by several richly-wooded islands amongst which are Lord's Island, St. Herbert's Island, Vicar's Island, and Ramps Holme. Lord's Island, the largest in the lake, situated perhaps a hundred yards from the shore, under Wallow Crag, was the strong-hold of the powerful family of the Ratcliffes, Earls of Derwent Water, whose possessions, it need hardly be said, were forfeited after the Rebellion of 1715, and transferred to Greenwich Hospital. On St. Herbert's Island are the remains of an Hermitage, said to have been fixed there by St. Herbert, the contemporary and friend of St. Cuthbert, in the seventh century.

"Stranger! not unmoved
Wilt thou behold this shapeless mass of stones,
The desolate ruin of St. Herbert's Cell.
Here stood his threshold; here was spread the roof,
That sheltered him, a self-secluded Man.
"When, with eye upraised
To Heaven, he knelt before the crucifix,

While o'er the Lake the Cataract of Lodore
Peal'd to his orisons, and when he paced
Along the beach of this small isle, and thought
Of his Companion, he would pray that both
(Now that their earthly duties were fulfilled)
Might die in the same moment.—Nor in vain
So prayed he—as our Chroniclers report,
Though here the Hermit number'd his last day,
Far from St. Cuthbert, his beloved Friend—
Those holy Men both died in the same hour.

There is also on this lake a FLOATING ISLAND, which is generally under water, but it occasionally rises to the surface for a short time, when it again sinks. The cause of this phenomenon has not been very clearly explained. The most probable supposition is that the mass is buoyed up, being swoln by gas produced by decomposed vegetable matter. On piercing it with a boat-hook, gas (*carburetted hydrogen and azote*) issues in abundance. The last appearance of this island was in the summer of 1842. The scenery of Derwent Water is distinguished for its wild sublimity and magnificence.

The Vale of Keswick stretches, without winding, nearly North and South, from the head of Derwent Water to the foot of Bassenthwaite Lake. It communicates with Borrowdale on the South; with the river Greta, and Thirlmere, on the East, with which the traveller has become acquainted on his way from Ambleside; and with the Vale of Newlands on the West—which last vale he may pass through in going to, or returning from, Buttermere. The best views of Keswick Lake are from Crow Park; Friar's Crag; the Stable-field, close by; the Vicarage; and from various points in taking the circuit of the lake. More distant and perhaps fully as interesting views, are from the side of Latrigg, from Ormathwaite, and thence along the road at the foot of Skiddaw towards Bassenthwaite, for about a quarter of a mile. There are fine bird's-eye views from the Castle hill; from Ashness, on the road to Watendlath; and by following the Watendlath stream down towards the cataract of Lodore. This lake also, if the weather be fine, ought to be circumnavigated. There are good views along the western side of Bassenthwaite Lake, and from Armathwaite at its foot; but the eastern side from the high road has little to recommend it. The traveller from Carlisle, approaching by way of Ireby, has, from the old road on the top of Bassenthwaite-hause, much the most striking view of the Plain and Lake of Bassenthwaite, flanked by Skiddaw, and terminated by Wallow Crag on the south-east of Derwent Lake; the same road commands an extensive view of the Solway Frith and the Scotch Mountains. They who take the circuit of Derwent Lake, may at the same time include BORROWDALE, going as far as Bowder-stone, or Rosthwaite. Borrowdale is also conveniently seen on the way to Wastdale over Sty Head; or, to Buttermere, by Seatoller and Honister Crag; or, going over the Stake, through Langdale, to Ambleside. Buttermere may be visited by a shorter way through Newlands, but though the descent upon the vale of Buttermere, by this approach, is very striking, as it also is to one entering by the head of the vale, under Honister Crag, yet, after all, the best entrance from Keswick is from the lower part of the vale, over Whinlatter to Scale Hill, where there is a roomy Inn, with very good accommodation.

Excursions from Keswick.

CASTLE HEAD.

CASTLE HEAD, or CASTLET, as it is called by the inhabitants, is considered the best Station in the neighbourhood (of easy access) for a bird's-eye view of the lake and surrounding mountains, and has consequently been selected for our Diagram. [*See Plate No.* 2.] Castle Head is approached by a good footpath, which strikes out of the Borrowdale road half a mile from Keswick, and leads by a winding ascent to the summit of the hill.

FRIAR'S CRAG

Is a rocky promontory which stretches out into the lake about one mile from Keswick, and, being the favourite promenade of the residents, is readily pointed out to strangers. From this Station nearly the whole circumference of the lake is viewed. After much rain the waters of Lodore may not only be seen but heard from Friar's Crag, and in the stillness of night the roar of this, combined with the murmur of other distant cataracts, has a solemn and soothing effect on the contemplative mind.

Plate 2

Mountains as seen from Castlehead, Keswick.

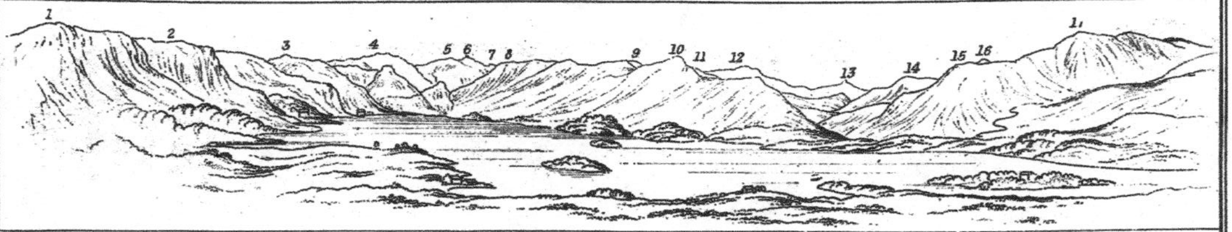

1 Wallow Crag
2 Falcon Crag
3 The Knotts
4 Glarramara
Brund (below)
5 Great End
Castle Crag (below)
6 Scawfell Pike
7 Scawfell
8 Gate Crag
9 Gold Scalp
10 Catbell
11 Robinson
12 High Stile
13 Red Pike
14 Knott Rigg
15 Rawling End
16 Knott Pike
17 Causey Pike

Mountains as seen from the foot of Dun Mallet on Ullswater.

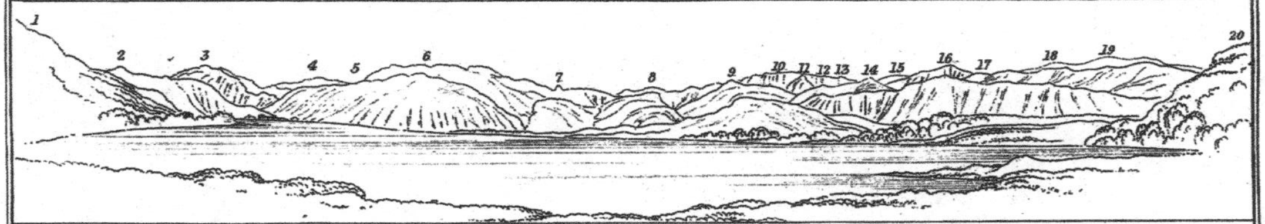

1 Swarth Fell
2 Stile End
3 Winter Crag
4 Dow Crag in Hartshop
5 Hallen Fell
6 Place Fell
7 Stone Cross Pike
8 Dolly Waggon Pike
Birk Fell (below)
9 High Spine How
The Knotts (below)
10 Hellvellyn
11 Cachety Cam
12 Glenridding Dodd
13 Hellvellyn Low Man
14 Herring Pike
15 Keppel Cove Head
16 Raise
17 Gowbarrow
18 Greenside
19 Glencoin Fell
20 Soulby Fell

J. Flintoft del.t W. H. Lizars sc.

General AQUATIC EXCURSION on DERWENT WATER.

1	Friar's Crag	1	1¼	St. Herbert's Island	4¾
¼	Lord's Island	1¼	¾	Water End Bay, with a little walking	5½
¼	Stable Hills	1½	1¼	Derwent Isle	6¾
¼	Broom Hill	1¾	¼	Strand's Piers	7
¾	Barrow Landing Place ...	2½	½	Keswick...	7½
¾	Floating Island	3¼			
¼	Mouth of the river Grange ...	3½			

Parties navigating the lake for the purpose of seeing its beauties, would do well to instruct the boatman to follow the directions pointed out in the above Table.

To BORROWDALE and round DERWENT WATER.

2	Barrow-house and Cascade ...	2	1	Return to Grange	6
1	Lodore* Waterfall	3	4	Portinscale	10
1	Grange	4	2	Keswick	12
1	Bowder Stone	5			

The scenes observable on this Excursion are viewed to the greatest advantage by commencing on the eastern, or Borrowdale road, having on the left Castle Head, and the broad fronts of Wallow Crag and Falcon Crag. A deep cleft in the face of Wallow Crag is visible from the road, which bears the name of the *Lady's Rake*, from the circumstance, it is said, of the Countess of Derwentwater having made her escape up this ravine when intelligence of her husband's arrest reached her. Two miles from Keswick is Barrow House, the seat of Joseph Pocklington Senhouse, Esq. It is surrounded by fine old trees, and has within the grounds a pretty cascade, which may be seen on application at the lodge. A mile more will bring the traveller to the celebrated FALL OF LODORE, which lies immediately at the back of the premises belonging to the inn. After incessant rains this Waterfall, with its accompaniments, is a noble object, but unfortunately for those who visit the Lakes, not one in a hundred sees it at such a time. The stream falls through a chasm between the two towering perpendicular rocks of Gowdar Crag upon the left, and Shepherd's Crag upon the right. These cliffs are most beautifully enriched with oak, ash, and birch trees, which fantastically impend from rocks where vegetation would seem almost impossible. The height of the fall is about 150 feet, and has been noticed by the late Dr. Southey in the following amusing lines:—

How does the water come down at Lodore?
Here it comes sparkling,
And there it lies darkling;
Here smoking and frothing,
Its tumult and wrath in.
It hastens along, conflictingly strong,
Now striking and raging, as if a war waging,
In caverns and rocks among.
Rising and leaping,
Sinking and creeping,
Swelling and flinging,
Showering and springing,
Eddying and whisking,
Spouting and frisking,
Turning and twisting
Around and around,
Collecting, disjecting,
With endless rebound,
Smiting and fighting,
A sight to delight in,
Confounding, astounding,
Dizzying and deafening the ear with its sound.
. . . .
And so never ending, but always descending,
Sounds and motions for ever and ever are blending
All at once and all o'er, with a mighty uproar—
And in this way the water comes down at Lodore.

At Lodore, in still weather, an extremely fine echo is to be heard, and a cannon is kept at the Inn to be discharged for the gratification of strangers. A mile from Lodore is the village of Grange, where there is a bridge that crosses the Derwent. Should the Tourist wish to see BOWDER STONE, the road into Borrowdale must be kept for one mile further. This stone is of prodigous bulk, and lies like a ship upon its keel.* It is 62 feet long and 36 feet high; its circumference is 84 feet, and it weighs about 1771 tons, This massive body has, probably by some great convulsion of nature, been detached from the rock above; but that it should stop in this position, after the violence of its motion in its descent from the mountain, is surprising, for to place it in its present position, or even to move it by any power

* Mr. Wordsworth has thus described its peculiar position:
" Upon a semicirque of turf-clad ground
A mass of rock resembling as it lay,
Right at the foot of that moist precipice,
A stranded ship, with keel upturned, that rests
Careless of winds and waves."

of art, seems utterly impossible. From this point a fine view of the upper part of Borrowdale is obtained, with the village of Rosthwaite and Castle Crag on the right, Eagle Crag and Glaramara in front, and Scawfell Pikes in the extreme distance. Returning to Grange Bridge, cross it, and pass through the village of Grange to the hamlet of Manesty, near which place is a medicinal spring. Proceeding at a considerable height along the open side of Cat Bells, which commands one of the best views of the lake and valley, and soon crossing the broad opening of Newlands, the road enters the village of Portinscale, from which place it is one mile and a half to Keswick.

WATENDLATH.

2	Over Barrow Common ...	2
½	Ashness Bridge	2½
1¼	Wooden Bridge between High Lodore and Watendlath	3¾
1¼	Watendlath	5
2	Rosthwaite	7
6	Keswick, by Bowder Stone and Lodore	13

The valley of WATENDLATH is interesting for its seclusion and loneliness, and the primitive character of its inhabitants. It runs paralled with the Vale of Borrowdale on the east, and is not easily accessible except on foot or horseback. The stream which forms the waterfall at Lodore issues from a beautiful little circular lake situated in this upland valley. The road thither from Keswick turns from the road to Borrowdale beyond Wallow Crag, and passes just behind Barrow House. A pretty rustic bridge crosses the stream where it issues from the tarn, and leads over the Borrowdale fells to Rosthwaite, a little above Bowder Stone. "This is a very pleasant morning's ride from Keswick; it may be varied on foot by turning to the left instead of the right at Watendlath, and crossing the Wythburn fells to Thirlmere, distant about four miles from Watendlath, over rough, heathery, trackless hills, which, on a fine day, especially when the heath is in blossom, form a wild and delightful walk."

Watendlath may also be visited on foot by High Lodore. The road turns off at the first house beyond the Inn, and is very steep till the stream is gained. A deviation to the left will presently unfold a truly magnificent view of the lake and the Skiddaw rage, through the deep chasm of the waterfall. From this place it is half a mile to the wooden bridge before alluded to.

VALE OF ST. JOHN.

From Keswick through the secluded VALE of ST. JOHN is an interesting excursion of about thirteen miles. A visit to the DRUID'S TEMPLE may be included in this walk by pursuing the old road to Penrith, which strikes off to the right about a quarter of a mile from the toll-bar. The Circle is a mile and three-quarters from Keswick, and will be found in a field on the right of the road, and just on the crown of the hill, whence there is a commanding view of Saddleback, Skiddaw, Helvellyn, and many of the highest mountains in Cumberland. The stones that form this Temple are forty-eight in number, describing a circle of near a hundred feet in diameter. Most of these stones are a species of granite, and all of them varying in form and size. On the eastern side of this monument there is a small inclosure formed within the circle by ten stones, making an oblong square, seven paces in length and three in width, which recess Mr. Pennant supposes to have been alloted to the priest, a sort of *holy place*, where they met, separated from the vulgar, to perform their rites and divinations, or to sit in council to determine on controversies, or for the trial of criminals. Within a short distance from Threlkeld, four miles from Keswick, a road branches off to the right to the Vale of St. John, "a very narrow dell, hemmed in by mountains, through which a small brook makes many meanderings, washing little inclosures of grass-ground which stretch up the rising of the hills. A nearer bridle-road into the vale leaves the Penrith road at the third milestone. In the widest part of the dale you are struck with the appearence of an ancient ruined castle, which seems to stand upon the summit of a little mount, the mountains around forming an amphitheatre. * * * As you draw near, it changes it figure, and proves no other than a shaken massive pile of rocks which stand in the midst of this vale, disunited from the adjoining mountains, and have so much the real form and resemblance of a castle, that they bear the name of the CASTLE ROCKS OF ST. JOHN." This is the scene of Sir Walter Scott's Poem of *The Bridal of Triermain.* The Tourist, after leaving the vale, enters the high road from Ambleside to Keswick, four miles and a half from the latter place, which road he must pursue in returning to his inn.

KESWICK to STY HEAD.

4	Grange Bridge...	4
1	Bowder Stone	5
1	Rosthwaite	6
½	Burthwaite Bridge	6½
½	Strand's Bridge	7
½	Seatoller Bridge	7½
½	Seathwaite Bridge	8
½	Seathwaite, which is opposite the Black Lead Mines...	8½
1	Stockley Bridge	9½
1¾	Sty Head Tarn...	11¼
¾	Sty Head	12
12	Back to Keswick	24

This road, as far as Bowder Stone, has already been noticed. A little beyond Bowder Stone, in the gorge of Borrowdale, rises a high and nearly detached rock called Castle Crag, the site of an ancient fortification, supposed to be of Roman origin, and to have been used to guard the Pass and secure the treasures contained in the bosom of these mountains. The Saxons, and, after them, the Furness monks, maintained the fort for the same purpose. All Borrowdale was given to the monks of Fnrness, probably by one of the Derwent family, and Adam de Derwentwater gave them free ingress and egress through all his lands. The Grange was the place where they laid up their grain and their tithe, and also the salt they made at the Salt Spring, of which works there are still some vestiges remaining below Grange. From the summit of this rock the views are so extensive and pleasing that they ought not to be omitted. "Beyond the hamlet of Rosthwaite (where there is a small public-house, the last in the valley), six miles from Keswick, the valley divides into two branches, that to the left being called Stonethwaite, and that on the right Seathwaite. Stonethwaite is subdivided into two branches, of which the eastern, called Greenup, leads into the fells towards the head of Easedale, and so communicates with Grasmere; while the Langstreth branch turns south, and communicates with Langdale by the Pass of the Stake. On entering Stonethwaite, Eagle Crag is a prominent object. Following the valley of Seathwaite, which is the principal vale, we come, two miles from Rosthwaite, to a large substantial farmhouse, called Seatoller, near which a rough mountain-road diverges to the right, and, passing under Honister Crag, descends upon Buttermere. A mile beyond Seatoller the Black-lead (or as it is provincially termed 'Wad') mine indicates its position, high on the hill-side, by those unsightly heaps of rubbish which always attend mining operations. Under the mine, and rather nearer to Seatoller, a dark spot is seen in the copse-wood, which thus far clothes the hill. These are the celebrated Borrowdale Yews, four in number, besides some smaller ones. Among them one is prominent, which, being in the vigour of its age, and undecayed, ranks among the finest specimens of its kind in England. This tree is seven yards in circumference at the height of four feet from the ground. The Lorton Yew is larger and that in Patterdale Church-yard may have equalled or exceeded this in size, but they have lost the mighty limbs and dark umbrageous foliage, contrasting so well with the rich chesnut-coloured trunk, which are here still to be seen in mature perfection. Mr. Wordsworth, after commemorating that of Lorton, continues—

> Worthier still of note
> Are those fraternal Four of Borrowdale,
> Join'd in one solemn and capacious grove;
> Huge trunks!—and each particular trunk a growth
> Of intertwisted fibres serpentines,
> Up-coiling, and inveterately convolved,—
> Nor uniform'd with Phantasy, and looks
> That threaten the prophane;—a pillar'd shade,
> Upon whose grassless floor of red-brown hue,
> By sheddings from the pining umbrage tinged
> Perennially—beneath whose sable roof
> Of boughs, as if for festal purpose, deck'd
> With unrejoicing berries, ghostly Shapes
> May meet at noontide—Fear and trembling Hope,
> Silence and Foresight—Death the Skeleton,
> And Time the Shadow,—there to celebrate
> As in a natural temple scatter'd o'er
> With altars undisturb'd of mossy stone,
> United worship; or in mute repose
> To lie, and listen to the mountain-flood
> Murmuring from Glaramara's* inmost caves.

"At the hamlet of Seathwaite, wood and cultivation end. There is no inn at Seathwaite, but the tourist will find ample refreshments at Mrs. Dixon's, a private house in the village. The road, now reduced to a horse-track, follows the rapidly-ascending bed of the stream for a mile further, and then, turning sharp over a little bridge, thrown across that branch of the Grange river which comes down from Esk Hause, begins im-

* A part of the Borrowdale Fells, above Rosthwaite, between Seathwaite and Langstreth.

mediately to mount Sty Head. But Stockley Bridge, as it is called, will detain our attention for a time, as a perfect miniature model of a bridge and waterfall. It is a rough stone arch apparently wedged rather than cemented together, hardly two yards in span, or one in breadth, with no parapet except a slight elevation of the outer stones on either side, between which there seems hardly room for a horse to plant his feet. It is thrown over a rocky cleft, ten or twelve feet above the stream, with a small glittering cascade above, and a sea-green pool below; for the purest spring is not more free from taint of moss than the water which descends from these hills. Small as it is, this is one of the most perfect specimens left of those *native* bridges, the gradual disappearance of which is generally regretted.*

"The height of Sty Head above the valley is said by Mr. Baines (' *Companion to the Lakes*') to be 1250 feet; this, however, is its height above the sea: its height above Stockley Bridge probably does not exceed 750 or 800 feet. At the top of the first ascent is a small plain, in which lies a narrow sheet of water, called Sty Head Tarn. Beyond it, the road still rises, until turning a sharp point of a rock, with a chasm at our feet, Wastdale lies in view more than a thousand feet below; while in front the precipices of the Pikes rise double that height The grandeur of the scene is enhanced by the suddenness with which it comes into view. On the Wastdale side of the Gable, garnets abound in the hard flinty slate. Sty Head Tarn is fed by a rill from Sprinkling Tarn, the source of one branch of the Grange river, which lies some hundred feet higher, under the broad front of Great End. Horses may be taken in the ascent of the Pikes to Sprinkling Tarn, or, with care, even to Esk Hause. Passing south of the Tarn, we proceed eastward up the hill side towards Esk Hause, where this route unites with the shorter and more direct one, which follows the water up from Stockley Bridge."

The return to Keswick may be varied, by striking over the mountains into the Vale of Langstreth and through Stonethwaite.

* The character of this bridge has been lamentably changed since this description of it was written. The bridge itself has been made wider by two or three feet, and the former singularly picturesque appearance of the parapet has been completely destroyed by the introduction of an unsightly smooth coping.

ASCENT OF SCAWFELL.

The last Excursion conducted the tourist to Sty Head and as far as Esk Hause, in the ascent of Scawfell. The present will place him on the summit of the highest mountain in England. The following account of a visit to this lofty eminence is extracted from a letter by a friend of Mr. Wordsworth, and may not be uninteresting.

"Having left Rosthwaite in Borrowdale, on a bright morning in the first week of October, we ascended from Seathwaite to the top of the ridge called Esk Hause, and thence beheld three distinct views:—on one side, the continuous Vale of Borrowdale, Keswick, and Bassenthwaite,—with Skiddaw, Helvellyn, Saddleback, and numerous other mountains,—and, in the distance, the Solway Frith and the Mountains of Scotland;—on the other side, and below us, the Langdale Pikes—their own vale below *them*;—Windermere,—and, far beyond Windermere, Ingleborough in Yorkshire. But how shall I speak of the deliciousness of the third prospect! At this time, *that* was most favoured by sunshine and shade. The green Vale of Esk—deep and green, with its glittering serpent stream, lay below us; and, on we looked to the Mountains near the Sea,—Black Comb pre-eminent, —and, still beyond, to the Sea itself, in dazzling brightness. Turning round, we saw the Mountains of Wastdale in tumult; to our right, Great Gable, the loftiest, a distinct, and *huge* form, though the middle of the mountain was, to our eyes, as its base.

We had attained the object of this journey; but our ambition now mounted higher. We saw the summit of Scawfell, apparently very near to us; and we shaped our course towards it; but, discovering that it could not be reached without first making a considerable descent, we resolved, instead, to aim at another point of the same mountain, called the *Pikes*, which I have since found has been estimated as higher than the summit bearing the name of Scawfell Head, where the Stone Man is built.

The sun had never once been overshadowed by a cloud during the whole of our progress from the centre of Borrowdale. On the summit of the Pike, which we gained after much toil, though without difficulty, there was not a breath of air to stir even the papers containing our refreshment, as they lay spread out upon a rock. The stillness seemed to be not of this world:—we paused, and kept silence to listen; and no sound could be heard: the Scawfell Cataracts were voiceless to us; and there was not an insect to hum in the air. The vales which we had seen from Esk Hause lay yet in view; and, side by side with Eskdale, we now saw the sister vale of Donnerdale terminated by the Duddon Sands. But the majesty of the mountains below, and close to us, is not to be conceived. We now beheld the whole mass of Great Gable from its base,—the Den of Wastdale at our feet—a gulph immeasurable: Grasmoor and the other mountains of Crummock—Ennerdale and its mountains: and the Sea beyond! We sat down to our repast, and gladly would we have tempered our beverage (for there was no spring or well near us) with such a supply of delicious water as we might have procured, had we been on the rival summit of Great Gable; for on its highest point is a small triangular receptacle in the native rock, which, the shepherds say, is never dry.* There we might have slaked our thirst plenteously with a pure and celestial liquid, for the cup or basin, it appears has no other feeder than the dews of heaven, the showers, the vapours, the hoar frost, and the spotless snow.

While we were gazing around, "Look," I exclaimed, "at yon ship upon the glittering sea!" "Is it a ship?" replied our shepherd-guide. "It can be nothing else," interposed my companion; "I cannot be mistaken, I am so accustomed to the appearance of ships at sea." The Guide dropped the argument; but, before a minute was gone, he quietly said, "Now look at your ship; it is changed into a horse." So it was,—a horse with a gallant neck and head. We laughed heartily; and, I hope, when again inclined to be positive, I may remember the ship and the horse upon the glittering sea; and the calm confidence, yet submissiveness, of our wise Man of the Mountains, who certainly had more knowledge of clouds than we, whatever might be our knowledge of ships.

* This natural basin was reported to have been destroyed by the officers employed by Government on the Ordnance Survey, but the writer of this note has the satisfaction to state that when he ascended the Gable, in September, 1842, he found it uninjured, and full of water, although more than half covered by a Stone Man that had been erected on the summit of the mountain.—We may observe, once for all, that the term "Man" is provincially applied to the piles of stones erected on the tops of most of the lake hills and mountains.

I know not how long we might have remained on the summit of the Pike, without a thought of moving, had not our Guide warned us that we must not linger; for a storm was coming. We looked in vain to espy the signs of it. Mountains, vales, and sea were touched with the clear light of the sun. "It is there," said he, pointing to the sea beyond Whitehaven, and there we perceived a light vapour unnoticeable but by a shepherd accustomed to watch all mountain bodings. We gazed around again, and yet again, unwilling to lose the remembrance of what lay before us in lofty solitude; and then prepared to depart. Meanwhile the air changed to cold, and we saw that tiny vapour swelled into mighty masses of cloud, which came boiling over the mountains. Great Gable, Helvellyn, and Skiddaw were wrapped in storm; yet Langdale, and the mountains in that quarter, remained all bright in sunshine. Soon the storm reached us: we sheltered under a crag; and, almost as rapidly as it had come, it passed away, and left us free to observe the struggles of gloom and sunshine in other quarters. Langdale now had its share, and the Pikes of Langdale were decorated by two splendid rainbows. Before we again reached Esk Hause every cloud had vanished from every summit.

I ought to have mentioned, that round the top of Scawfell PIKE not a blade of grass is to be seen. Cushions or tufts of moss, parched and brown, appear between the huge blocks and stones that lie in heaps on all sides to a great distance, like skeletons or bones of the earth not needed at the creation, and there left to be covered with never-dying lichens, which the clouds and dews nourish; and adorned with colours of vivid and exquisite beauty. Flowers, the most brilliant feathers, and even gems, scarcely surpass in colouring some of those masses of stone which no human eye beholds, except the shepherd or traveller be led thither by curiosity; and how seldom must this happen! For the other eminence is the one visited by the adventurous stranger; and the shepherd has no inducement to ascend the PIKE in quest of his sheep; no food being *there* to tempt them.

We certainly were singularly favoured in the weather; for when we were seated on the summit, our conductor, turning his eyes thoughtfully round, said, "I do not know that in my whole life I was ever, at any season of the year, so high upon the mountains on so *calm* a day." (It was the 7th of October.) Afterwards we had a spectacle of the grandeur of earth and heaven commingled; yet without terror. We knew that the storm would pass away—for so our prophetic guide had assured us.

Before we reached Seathwaite, in Borrowdale, a few stars had appeared, and we pursued our way down the vale, to Rosthwaite, by moonlight.

If the tourist be bound from the Pikes into Eskdale, a direct and practicable, but somewhat difficult, descent may be found by way of Mickledore, a deep chasm separating Scawfell from the Pikes, at the bottom of which a narrow ridge, like the roof of a house, slopes into Eskdale on one side, and into Wastdale on the other. But the descent of Scawfell from this point ought not to be undertaken without a Guide well acquainted with the practicable passes of this mountain. It is encompassed by precipices and narrow terraces of turf and slanting sheets of naked rock; and a stranger might chance to find himself entrapped into some place, where to go backwards or forwards would be equally difficult and dangerous.

A tolerably straight course may be shaped from the Pikes into Wastdale down the breast of Lingmell, or, if the traveller be returning to Keswick, he may descend to Sty Head by the western side of the mountain, leaving Great End to the right, and keeping farther down the hill-side than would at first seem necessary, to avoid some deep and apparently impassable ravines, which run out from among the crags of Great End. These oblige him to descend below the level of Sty Head.

From Esk Hause an hour well used will take the walker, in a different direction, to the head of Langdale. The way lies past Angle Tarn, under the northern precipice of Bowfell. The best descent into Langdale is down a steep rugged gully, called Rosset Gill. The circuit from Keswick to Ambleside by Sty Head, the Pikes, Esk Hause, and Langdale, may be reckoned at thirty miles, and lies throughout among the finest scenery in the country.

SKIDDAW.

Skiddaw is the fourth English mountain in height, being 3022 feet above the level of the sea, and 2911 above Derwent Water. To the highest point from Keswick it is six miles, and is so easy of access that persons may ride to the summit on horseback. The approach to Skiddaw is by the Penrith road for about half a mile, chiefly along the banks of the Greta, to a bridge near the toll-bar. Having crossed the bridge, the road ascends somewhat steeply, and after passing Greta Bank skirts Latrigg at a considerable elevation. A little beyond the plantation the tourist will see another road, which he must take, though only for a few yards, when he must again turn, just beyond a gate on the left, at right angles, by the side of a fence to a hollow at the foot of the steepest hill in the ascent. From this place the road rises precipitously for almost a mile by the side of a stone wall, which it crosses about one-third of the way up, and then leaves on the right. The ascent then becomes easy over a barren moor, called Skiddaw Forest, to the foot of the low Man, where there is a fine spring of water. Beyond this well, having the first and second summits, or *Men*, as they are called, on the left, the road ascends easily by a good beaten track to the third Man, which is the highest point that can be seen from the valley, and from this elevated station the whole extent of the vale beneath is most beautifully displayed. After passing the fourth and fifth heap of stones, the traveller will soon place himself upon the highest summit of this mountain. Derwent Water cannot be seen from this lofty eminence, being obscured by others of less elevation, which hide also the high grounds lying between Wythburn and Langdale. On the right of the third Man appears a most magnificent assemblage of mountains. In a south-western direction, is seen that sublime chain extending from Coniston to Ennerdale, amongst which Scawfell stands pre-eminent, having on its left Great End, Hanging Knott, Bowfell, and the fells of Coniston; and on the right Lingmell Crags, Great Gable, Kirkfell, Black Sail, the Pillar, the Steeple, and the Hay Cock, with Yewbarrow and part of the Screes through Black Sail. Black Comb may bs descried through an opening between the Gable and Kirkfell. To the north of the Ennerdale mountains are those of Buttermere;

and High Crag, High Stile, and Red Pike peer nobly over Cat Bells, Robinson, and Hindscarth. Still further to the north, rising from the vale of Newlands, is Rawling End, whence, aspiring, are Cawsey Pike, Scar Crag Top, Sail, Ill Crags, Grasmire, and Grisedale Pike. On the right of Grisedale Pike and Hobcarten Crag is Low Fell, over which, in a clear atmosphere, may be observed the northern part of the Isle of Man; and, perhaps one day out of a hundred, Ireland may be seen. The town and Castle of Cockermouth are distinctly seen over the foot of Bassenthwaite, with Workington at the outlet of the Derwent on its left. Whitehaven is hid from our view, but all the sea coast from St. Bees' Head by Solway Frith to Rockliff Marsh may be easily traced. Over the northern end of Skiddaw, Carlisle, if the state of the atmosphere be favourable, may be plainly seen, and the Scotch mountains of Criffel, &c., give a fine finish to the fertile plains of Cumberland. Eastward, Penrith and its Beacon are visible, with Cross-fell in the distance; and far away to the south-east the broad head of Ingleborough towers over the Westmorland fells. Saddleback here displays its pointed top, and nearly due south is seen the lofty summit of Helvellyn. The estuaries of the Kent and the Leven, separated by a hill called Yewbarrow, near Grange, are visible through the gap of Dunmail Raise; and Lancaster Castle may sometimes be seen beyond Gummershow at the foot of Windermere, with the aid of a telescope; but no part of the lake of Windermere can, as has been frequently stated, be discerned from this point.

The descent, for the sake of variety, might be made into the valley of Bassenthwaite, where refreshments may be had at the Castle Inn, near the foot of the lake, whence it is eight miles to Keswick by the eastern, and ten by the western road.

SADDLEBACK.

Saddleback is, in the opinion of some tourists, more worthy of a visit than Skiddaw. "Derwent Water," says Dr. Southey, "as seen from the top of Saddleback, is one of the finest mountain scenes in the country. The tourist who would enjoy it should proceed about six miles along the Penrith road, then take the road which leads to Hesket New Market, and presently ascend by a green shepherds' path which winds up the side of a ravine; and, having gained the top, keep along the summit, leaving Threlkeld Tarn below him on the right, and descend upon the Glenderaterra, the stream which comes down between Saddleback and Skiddaw, and falls into the Greta about two miles from Keswick." The ancient name of this mountain is Blencathra. The modern one of Saddleback has been given to it from the peculiarity of its formation, as seen from the neighbourhood of Penrith, where it takes something of the shape of a saddle. Its height is 2787 feet. At the base of an enormous perpendicular rock called Tarn Crag, near Linthwaite Pike, is Scales Tarn, a small lake deeply seated among the crags, which, from the peculiarity of its situation, is said to reflect the stars at noon-day. In Bowscale fell, and lying about three miles from Scales Tarn, in a north-easterly direction, is Bowscale Tarn, which sends a tributary to the Caldew. This tarn is the seat of a singular superstition, being supposed by the country people to be inhabited by two immortal fish; but we are not told in what way the belief originated.

"—— Both the undying fish that swim
In Bowscale Tarn did wait on him;
The pair were servants of his eye
In their immortality;
They moved about in open sight,
To and fro for his delight."

Song at the Feast of Brougham Castle.

GRISEDALE PIKE

Rises to the height of 2580 feet above the level of the sea. It is situated to the west of Keswick, above the village of Braithwaite, and well deserves a visit. Lovers of wild scenery will find much pleasure in continuing their walk along the ridge which connects Grisedale Pike with Grasmoor, returning by a pleasant morning's walk to Keswick over Causeway Pike.

RIDE from KESWICK to BUTTERMERE, through NEWLANDS.

1	Portinscale	1	1	Aikin	6½
2	Swinside	3	2	Newlands Haws	8½
1½	Stoneycroft, right	4½	1½	Inn at Buttermere	10
1	Bridge near Mill Dam ...	5½			

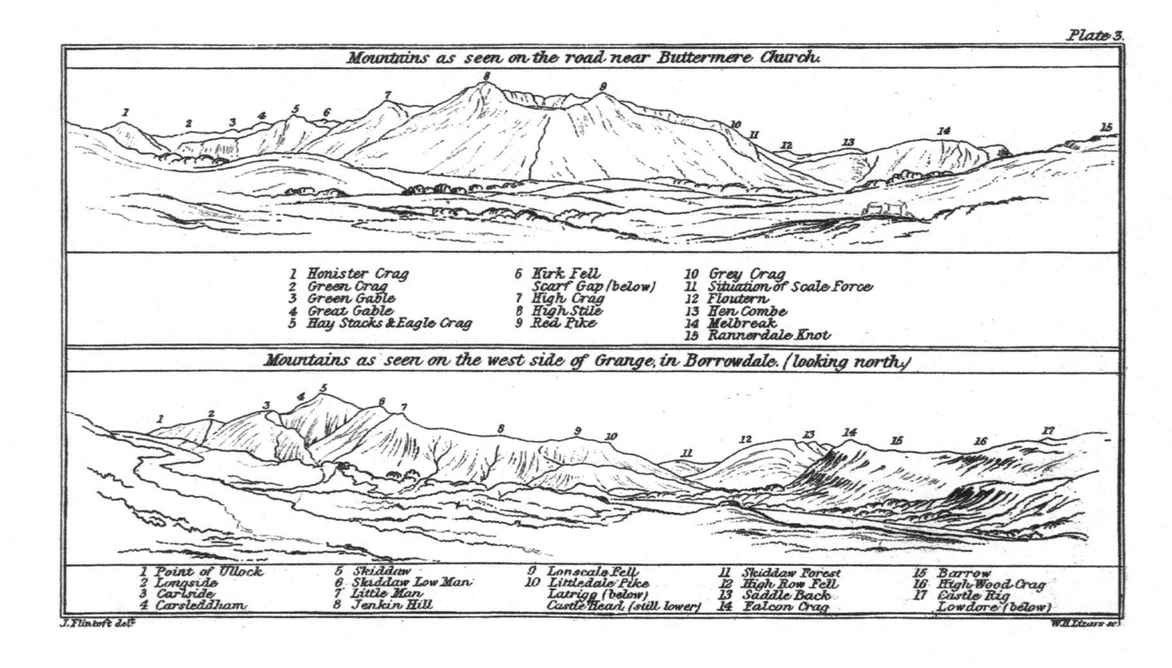
Plate 3.
Mountains as seen on the road near Buttermere Church.
1 Honister Crag
2 Green Crag
3 Green Gable
4 Great Gable
5 Hay Stacks & Eagle Crag
6 Kirk Fell
Scarf Gap (below)
7 High Crag
8 High Stile
9 Red Pike
10 Grey Crag
11 Situation of Scale Force
12 Floutern
13 Hen Combe
14 Melbreak
15 Rannerdale Knot
Mountains as seen on the west side of Grange, in Borrowdale. (looking north)
1 Point of Ullock
2 Longside
3 Carlside
4 Carsleddham
5 Skiddaw
6 Skiddaw Low Man
7 Little Man
8 Jenkin Hill
9 Lonscale Fell
10 Littledale Pike
Latrigg (below)
Castle Head (still lower)
11 Skiddaw Forest
12 High Row Fell
13 Saddle Back
14 Falcon Crag
15 Barrow
16 High Wood Crag
17 Castle Rig
Lowdore (below)
J. Flintoft delt
W.H. Lizars sc

The road to Newlands is by the village of Portinscale, and thence between Foe Park Woods and Swinside, to the Three Road Ends. The one on the right, skirting the southern flank of Swinside for some distance, leads through Newlands to Buttermere. At Rawling End (a mountain so called) the scenery is excellent, either looking back in the direction of Skiddaw, across the valley towards Cat Bells, or up the vale of Newlands. A fine branch of the vale of Newlands extends from Emerald Bank to Dale Head, guarded on the south by Maiden Moor and High Crag, and on the north by Goldscope* and Hindscarth. Above Keskadale, the last houses in the valley, the road ascends steeply to Newlands Haws, through the sides of which Great Robinson is advantageously seen. In the descent from the Haws to Buttermere the road runs at an alarming height above the ravine which separates this from the opposite hill called Whitelees. The chain of mountains developed in the descent of the Haws is the most magnificent in the whole circumference of the valley. The appearance of High Stile and of the whole visible horizon from Green Crags to Red Pike is scarcely equalled in Cumberland. (*See Plate No.* 3). The white stream called Sourmilk Gill, issuing from Bleaberry Tarn, or Burtness Tarn, down the rocky steep, forms a beautiful feature in the landscape. The road passes a neat little chapel recently erected by the Rev. Mr. Thomas on the site of a still smaller one, which was said to have been the smallest in England, and not capable of containing within its walls more than half a dozen households. At a short distance from the chapel stands the Inn where Mary Robinson, the Beauty of Buttermere, was for a number of years the unceasing object of public curiosity.

THE LAKE OF BUTTERMERE

Is one mile and a quarter in length, and little more than half a mile in breadth. Buttermere Moss and Great Robinson bound it on the east; Hay Stacks, so called from their form, High Crag, High Stile, and Red Pike rising to a great height, enclose it on the west; whilst Fleetwith and Honister Crag, at the head of the

* Probably so called from the quantity of gold and silver yielded by the Copper and Lead Mines worked here in the time of Queen Elizabeth.

lake, seem to shut out all communication southwards. At the north end, or outlet of the lake, it is separated from Crummock Water by meadows and luxuriant woods and hedge-rows, over which is seen at some distance, Lowfell, an eminence which separates Lowes Water from Lorton. Buttermere affords excellent sport for the angler.

Most persons content themselves with what they can see of Buttermere in one day, but many days might be profitably employed in exploring the beauties of this secluded vale. To such transient visitors it is recommended to see SCALE FORCE, one of the highest waterfalls in the country. The road to this place is by a footpath across the fields, which, from the soft and boggy nature of the ground, is anything but agreeable in damp weather; a better arrangement will therefore be, to take a boat at the head of Crummock Water, and proceed to the stream which issues from the fall, where parties are usually landed. From this point it is a mile to the Force, which is one clear fall of 160 feet between two vast perpendicular walls of syenite, beautifully adorned with numerous small trees which grow in the fissures of the rock, and are nourished by the spray of the falling waters. On returning to the boat, row direct to Ling Crag, a little rocky promontory at the foot of Melbreak, and from a point two or three hundred yards above this promontory is the best Station for a view of the two lakes of Crummock and Buttermere, and the surrounding mountains.

CRUMMOCK WATER

Is bounded on the east by the lofty mountain of Whiteside, Grassmoor, and Whitelees; and Melbreak is the western barrier for a considerable distance. Scale Hill is upwards of three miles from Ling Crag, and, if time should permit, parties may resort thither for refreshment at an excellent inn, and afterward return to Buttermere. The road recommended in the return to Keswick is by Borrowdale.—A mile and a half from the Inn at Buttermere, Hassness, the residence of — Benson, Esq., is passed on the right, and half a mile more will bring the traveller to a farmhouse called Gatesgarth.

[From this place a mountain-road strikes off to the right, between Haystacks and High Crag, to Ennerdale (six miles), by

Plate 4

Mountains as seen from Whinlatter.

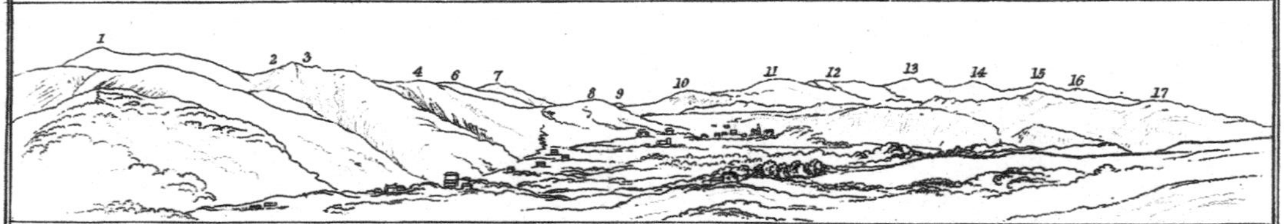

1 Skiddaw
Carlside (below)
Dodd (still lower)
2 Carsleddham (below)
3 Skiddaw Low Man
4 Jenkin Hill
6 Lonscale Fell
7 Saddleback
8 Latrigg
9 Little Melfell
10 Setnalianing
11 Great Dod
St. John's Ridge (below)
12 Watson Dod
Wallow Crag (below)
13 Stybarrow Dod
14 Whiteside
15 Helvellyn Low Man
Bleaberry Fell (below)
Falcon Crag (Still lower)
16 Eagle Crag
17 High Seat

Mountains as seen from the Road from Scale Hill to Loweswater, near the Mile Post.

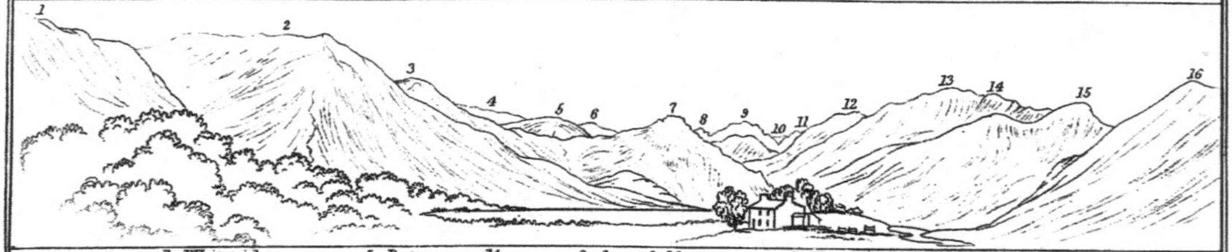

1 Whiteside
2 Grassmoor
3 Whitless Pike
4 Robinson
5 Buttermere Moss
6 Honister Crag
7 Rannerdale Knot
8 Green Gable
9 Great Gable
Haystacks (below)
10 Scawfell Middle Pike
Scarf gap (below)
11 Kirk Fell
12 High Crag
13 High Stile
14 Bleaberry Tarn (below)
15 Red Pike
16 Melbreak

the Pass of Scarf Gap, and is met by another path over Black Sail, on the opposite side of the valley of Gillerthwaite, which descends through the Vale of Mosedale, between Kirkfell and the Pillar to Wastdale Head (six miles). These roads are indicated on the Map. A horse may be taken over these hills in dry weather, but those who can bear walking will find it much pleasanter than riding: indeed much of the road *must* be passed on foot. It will be prudent to take a guide.]

From Gatesgarth the road to Borrowdale is by a laborious ascent of nearly three miles to the summit of Buttermere Haws, having the almost perpendicular rock of Honister Crag on the right and Yew Crag on the left hand. In both these there are extensive quarries of valuable roofing slate. A very interesting combination of mountains is exhibited from the top of the road, which begins to descend rapidly to Seatoller, in Borrowdale, from whence it is a mile and three-quarters to Rosthwaite, where there is a public-house. From thence, passing Bowder Stone, Grange (*where consult Diagrams, Plate* 3), and Lodore, it is six miles to Keswick. This Excursion may be made (but with some difficulty) in a car.

DRIVE to SCALE HILL at the Foot of CRUMMOCK WATER, and BUTTERMERE by WHINLATTER.

2½	Braithwaite	2½	4	Scale Hill	12
2½	Summit of Whinlatter	5	4	Buttermere	16
3	Lorton	8	9	Through Newlands to Keswick	25

The best approach to Crummock and Buttermere is by Whinlatter and Swinside to Scale Hill, ten miles, or by a more circuitous road through the Vale of Lorton, twelve miles. The road to Scale Hill leaves that to Bassenthwaite at the village of Braithwaite, where the ascent of Whinlatter commences, and although long and tedious, the Traveller is fully compensated for his toil by the noble retrospective views of the Vale of Keswick which are unfolded. *(See Diagrams, Plate* 4.) For two miles past the fourth milestone Grisedale Pike is on the left. A little beyond the sixth milestone, a road branches off to the left, along Swinside, and is the one which all persons whether on foot, on horseback, or even in carriages, should take, on their way to Scale Hill. On first entering this road the traveller may feel some disappointment, but, having ascended the hill, he will be charmed with the views of the Vale of Lorton, and the distant prospect of the Scotch mountains. The more circuitous route through the vale of Lorton turns off from the Cockermouth road at the Famous Yew Tree,* and joins the terrace-road just mentioned about a mile and a half from Scale Hill. A quarter of a mile beyond the junction of these roads, are two other roads; that on the left leads to Buttermere; the other to the Inn at Scale Hill.

Scale Hill is well situated for parties wishing to visit Crummock Water, Buttermere, Lowes Water, and Ennerdale.

From Scale Hill a pleasant walk may be taken to an eminence in Mr. Marshall's woods, and another, by crossing the bridge at the foot of the hill, upon which the Inn stands, and turning to the right, after the opposite hill has been ascended a little way, then following the road that leads towards Lorton for about half a mile, looking back upon Crummock Water, &c., between the openings of the fences. *(See Diagrams, Plate* 4.) Turn back and make your way to

LOWES WATER,

A small lake, about a mile in length, situated in a deep secluded valley about two miles from Crummock, and surrounded by the bold mountain of Blake Fell, Low Fell, and Melbreak. The valley is prettily wooded, and has an air of pastoral beauty. It is only seen to advantage from the other end, therefore any traveller approaching from the foot must look back upon it on arriving at its head.

* "——— pride of Lorton Vale,
Which to this day stands single, in the midst
Of its own darkness, as it stood of yore
Not loth to furnish weapons for the bands
Of Umfraville or Percy, ere they march'd
To Scotland's heath: or those that crossed the sea,
And drew their sounding bows at Azincour;
Perhaps at earlier Crecy, or Poictiers.
Of vast circumference, and gloom profound,
This solitary tree, a living thing,
Produced too slowly ever to decay;
Of form and aspect too magnificent
To be destroyed."

The following Table will shew the route to be observed in a

WALK round LOWES WATER from SCALE HILL.

¾	Lowes Water Church ...	¾
1½	Thence by Kirk Head, Bar Gate, Steel Bank, and High Nook, to Water Yeat	2¼
½	Gill falling from Carling Knott	2¾
1	Place, or High Water End	3¾
¾	Bottom, or Low Water End	4½
1	Crabtree Beck	5½
1	Join the road from Scale Hill to the Chapel at the Smithy	6½
½	Scale Hill	7

CRUMMOCK WATER AND BUTTERMERE

Are no where so impressive as from the bosom of Crummock Water. The following Excursion to Buttermere from Scale Hill will be found highly interesting.

LAND and WATER EXCURSION from SCALE HILL.

1	Boat House on Crummock Water	1
1½	Flat Fields at Rannerdale ...	2½
¾	Station above Ling Crag ...	3¼
1¾	To Scale Force and back... ...	5
1	Join the road at the head of the lake	6
1	Inn at Buttermere	7

ENNERDALE WATER

Is situated four miles to the south of Lowes Water. It is three-quarters of a mile in breadth, and extends two miles and a half in length. The scenery is wild and romantic, and beyond the head of the lake are seen some of the highest mountains in the country, of which the most conspicuous is the Pillar, rising to an elevation of 2893 feet.

> "It wears the shape
> Of a vast building, made of many crags;
> And in the midst is one particular rock,
> That rises like a column from the Vale,
> Whence, by our shephards, it is called THE PILLAR."

Owing to its difficulty of access to Southern Tourists, Ennerdale Water is rarely seen except from a distance. It may be approached from the Inn at Buttermere by Scale Force and Floutern Tarn; and also from Scale Hill through Mosedale* and by Floutern Tarn, and by several other mountain roads, all terminating at Crosdale, where the best views of the lake are obtained. There is a small public-house—the Boat House—at the foot of Ennerdale Lake, with a comfortable and pleasant sitting room, and plain accommodation for the night. The following Tables may be useful to the Traveller.

* This name is common to several valleys in the Lake District. It behoves Tourists to bear this in mind.

WALK from BUTTERMERE to its union with the Road from Crosdale to ENNERDALE WATER.

2	Sale Force...	2
2½	Floutern Tarn	4½
1½	Join the road from Crosdale to Ennerdale Water, where is one of the best views of the lake...	6
1	Ennerdale Water	7

Three roads on foot to CROSDALE, from SCALE HILL, by HIGH NOOK.

1	A mile on the high road to Lowes Water	1
¾	High Nook	1¾
3	Passage to Crosdale, over Blake Fell;	
	Or, to Crosdale, deviating at the top of Blake Fell on the left;	
	Or, to Crosdale by commencing the ascent with the rivulet on the left, at High Nook, and then turning on the right	4¾
1	Crosdale to Ennerdale Water (the finest views are halfway)	5¾

From SCALE HILL, by a Horse-road, to ENNERDALE WATER.

2¾	Lowes Water End, at the Head of Lowes Water	2¾
¾	Enter the Common	3½
1¾	Lampleugh Church	5¼
¾	Road on the left, beyond the Church	6
2¼	On this road by High Trees and Fell Dyke to Crosdale ...	8¼
½	Half way to the lake (the best prospect)...	8¾
½	Margin of the lake	9¼

From CROSDALE the Tourist may proceed to WASTDALE HEAD by pursuing the following route, or he may return to Buttermere by the foot-road over Scarf-gap after he has passed through the secluded valley of Gillerthwaite, as the upper part of Ennerdale is called. This road he will find marked upon the Map.

From CROSDALE, on foot, to the Eastern Side of ENNERDALE WATER, and through Ennerdale and Mosedale to WASTDALE HEAD.

1	Join the lake	1
½	Bowness	1½
2	Head of the lake...	3½
1½	Gillerthwaite	5
2½	Foot of the road to Buttermere over Scarf Gap	7½
½	Sheep-fold on the river side ...	8
¾	From which, with the stream on the left, ascend to the top of Black Sail	8¾
2¼	Wastdale Head, through Mosedale	11

TWO DAYS' EXCURSION TO WASTWATER.

Wast Water is seen to the greatest advantage on approaching it from the open country by the Strands at its foot, rather than by Sty Head. The latter road enters Wastdale at the head of the lake, and can only be taken on foot or on horseback. The Tourist, therefore, should commence this Excursion by going over Whinlatter to Scale Hill, already noticed, and proceeding by Lowes Water and Lampleugh Cross to Ennerdale Bridge, thence to Calder Bridge, from which place there is only one near road, and that is by Gosforth to the Strands in Nether Wastdale, near the foot of Wast Water. This road, although in part steep and not very good, may without difficulty be travelled over by light carriages ; but there is an excellent carriage road, which makes, however, a circuit of many miles, through Cockermouth, Workington, Whitehaven, and Egremont to Calder Bridge. By leaving Workington on the right, and passing from Cockermouth direct to Whitehaven the distance is shortened two miles.

From Scale Hill it is about two miles to Lowes Water; whence to Lampleugh Cross, where there are two small public houses, four miles; to Ennerdale Bridge, at the foot of Ennerdale, three miles more; and from Ennerdale Bridge seven miles to Calder Bridge, where excellent accommodation may be had at two comfortable Inns. The direct road from Ennerdale Bridge to Calder Bridge is over a dreary moor called Coldfell, and is extremely disagreeable and tiresome to drive over from the number of gates; so that it would be better to go by Egremont, although the distance would be increased four miles.

CALDER ABBEY.

Is one mile from Calder Bridge. Little of this ruin is left, but that little is well worthy of notice. It is situated on the north side of the river Calder, close to the residence of Captain Irwin, and was founded A. D. 1134 by the second Ranulph des Meschines for Cistercian monks, and was dependent on Furness Abbey.

From Calder Bridge to Gosforth, three miles; thence to the Strands public-house, four miles.

Circuitous Carriage Road.—This road, as far as the famous Lorton Yew-tree, eight miles from Keswick, has been already noticed. From the Yew-tree the turnpike-road must be kept, and after driving through a rich fertile country for four miles, the Traveller will reach

COCKERMOUTH,

A borough-town sending two members to Parliament, situate upon the Cocker, where it falls into the Derwent. Hats, coarse woollens, linen, and leather, are manufactured here. The Castle is for the most part in ruins, and belongs to General Windham, who occasionally resides there. Market on Monday and Saturday. Inns—*Globe, Sun.*

From Cockermouth to Whitehaven direct, is fourteen miles, and by Workington sixteen miles. On leaving Cockermouth, by turning aside a few steps, a fine view of the river Derwent and the Castle may be had from the bridge.

WORKINGTON

Is situated on the south bank of the Derwent, and has a good harbour well secured by a breakwater. In the vicinity of the town are several valuable coal mines, which are principally worked by Henry Curwen, Esq., the lord of the manor. Some of these were, a few years ago, destroyed by the sea breaking in upon them. The streets are irregularly built, but have of late years been much improved by modern erections. Workington Hall stands on a gentle eminence on the east side of the town, and is celebrated as having afforded an asylum to the unfortunate Mary Queen of Scots, after her escape from Dunbar Castle. Population, 7226.

WHITEHAVEN

Ranks the second town of importance in Cumberland. It is situated on a bay, and the harbour has been greatly improved by an elegant and substantial stone pier, said to be the largest in the kingdom. The town is built with great regularity, and the streets are spacious. The Castle is the residence of the Earl of Lonsdale, who is lord of the manor and proprietor of the coal mines, which perhaps are the most extraordinary in the world. In the William Pit there are 500 acres under the sea, and the distance is two miles and a half from the shaft to the extreme part of the workings. There is a stable also under the sea in this immense pit for forty-five horses. The shaft is 110 fathoms deep. The coals are principally exported to Ireland, and yield a large revenue to the noble proprietor. Ship-building is carried on here to some extent, and the principal manufactures of the town are linen sail-cloth, checks, ginghams, sheetings, thread, twine, cables, &c.

From Whitehaven it is six miles to EGREMONT by way of Hensingham, and seven by ST. BEES, "a place distinguished from very early times for its religious and scholastic foundations. 'St. Bees,' says Nicholson and Burn, 'had its name from Bega, a holy woman from Ireland, who is said to have founded here, about the year 650, a small monastery, where afterwards a church was built in memory of her. The aforesaid religious house having been destroyed by the Danes, was restored by William des Meschines, son of Ranulph, and brother of Ranulph des Meschines, first Earl of Cumberland after the conquest; and made a cell of a prior and six Benedictine monks to the Abbey of St. Mary at York.' After the dissolution of the monasteries, Archbishop Grindal founded a free school at St. Bees, from which the counties of Cumberland and Westmorland have derived great benefit; and recently, under the patronage of the Earl of Lonsdale, a college has been established there for the education of ministers of the English Church. The old Conventual Church was repaired under the superintendence of the Rev. Dr. Ainger, the late Head of the College; and is well worthy of being visited by any strangers who may be led to the neighbourhood of this celebrated spot." This collegiate institution is now in a highly flourishing condition, under the able management of the Rev. R. P. Buddicom.

EGREMONT

Is a neat little town, with about 1500 inhabitants, situate on the north side of the river Ehen, which flows from Ennerdale lake, seven miles distant. The road is good. The ruins of a Castle stand on an eminence to the west of the town. This fortress is not of very great extent, but bears singular marks of antiquity and strength.

From Egremont it is five miles of pleasant road to Calder Bridge, to which place the traveller was conducted by the route from Scale Hill.

Should the Tourist prefer the approach to Wast Water by Sty Head, the following is the route. The objects on the road have been described so far as Sty Head at p. 77, and the ascent of this mountain pass from the Strands is also described at p. 66.

First Day.—WAST WATER by Borrowdale, a Two Days' Excursion on horseback.

12	Sty Head	12
2	Wastdale Head	14
1	Head of Wast Water	15
½	Overbeck Bridge	15½
1	Netherbeck Bridge	16½
1¼	End of the direct road to Calder Bridge by Harrow Head	17¾
¾	Crook at the foot of the lake	18½
1½	Strands Public house	20
1½	Junction of the Strands road with the shortest road... ...	21½
2½	Gosforth	24
3	Calder Bridge, where there are two good Inns	27

Second Day.—See CALDER ABBEY, a mile from Calder Bridge, and then proceed

7	From Calder Bridge to Ennerdale Bridge	7
1½	Kirkland	8½
1	Road on the left to Egremont and Whitehaven	9½
½	Lampleugh Cross (the Cockermouth road is to the left) ...	10
1	Lampleugh Church	11
5	Scale Hill	16
11	Keswick over Swinside and Whinlatter	27

Round BASSENTHWAITE WATER.

8	Peel Wyke	8
1	Ouse Bridge...	9
1	Castle Inn	10
3	Bassenthwaite Sandbed	13
5	Keswick	18

Before bidding adieu to Keswick, the tour to Bassenthwaite Water should not be omitted. The lake of Bassenthwaite lies four miles north of Derwent Water, is four miles in length, and in some places near a mile in breadth. In commencing this Excursion proceed to the village of Braithwaite, at the foot of Whinlatter, which the tourist must leave on the left. Passing through the hamlet of Thornthwaite and skirting the base of the rugged mountains of Lord's Seat and Barf, the road undulates pleasantly through wood and glade on the margin of the lake, till it reaches Peel Wyke, where there is a small ale-house. A little beyond Peel Wyke the road turns off on the right at the guide-post to Ouse Bridge, which crosses the Derwent, where, and at Armathwaite close by, are the best views for those who keep the road generally pursued in making the circuit of the lake; but the pedestrian would be fully compensated if he were to deviate at the Castle Inn, one mile from Ouse Bridge, and follow the Hesket road for about a mile, and then turn on the right to the top of the Haws, from which is presented a magnificent view of Bassenthwaite and the Vales of Embleton and Isell. The distance from the Castle Inn to Keswick is eight miles; the road winds agreeably on the eastern side of the lake.

Plate 5

Mountains as seen from the Matterdale Road in Gowbarrow Park.

1 Swarth Fell
2 Hallen Fell (below)
3 Winter Crag Martindale
4 Place Fell
Birk Fell (below)
5 Red Screes
Deepdale Park (below)
Bleas (still lower)
6 Scandale Head
7 Dove Crag in Hartstop
8 Birks
Hall Bank (below)
9 Rydal Head
10 St. Sunday's Crag
11 Fairfield
12 Glenridding Dod
13 Dolly Wagon Pike
14 Bleaberry Fell
15 Stridding Edge
16 Herring Pike

Mountains as seen from Milking Hill, between the Inn and Blowick, in Patterdale.

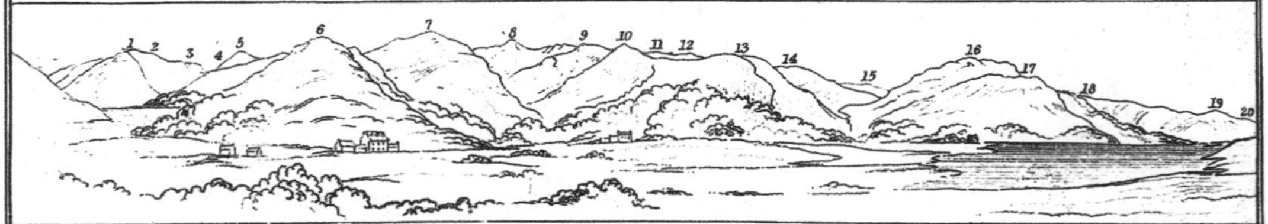

1 Hartstop Dod
2 Cawdale Moor
3 Woundale Head
4 Kirkstone Pass
5 Red Screes
Deepdale Park (below)
6 Blease
7 Birks
8 Dolly Wagon Pike
9 Eagle Crag
10 Striding Edge
11 Path to Helvellyn (below)
12 Helvellyn
13 Bleaberry Fell & Crag
14 Raise
15 Greenside
16 Herring Pike
17 Glenridding Dod
18 Glencoin Fell
19 Stinted Common Fell
20 Gowbarrow Park

J. Flintoft del.

W. H. Lizars sc.

ULLSWATER.

8 Moor End	8	*Patterdale to Penrith.*	
7 Gowbarrow Park	15	10 Pooley Bridge	10
5 Patterdale	20	6 Penrith	16

Ullswater is of an irregular figure, somewhat resembling the letter Z, and composed of three unequal reaches, the middle of which is somewhat longer than the northern one. The shortest is seen from the Inn at Patterdale, and is not half the length of either of the others. Ullswater is less than Windermere, but larger than the rest of the English lakes, and lies engulphed in the majestic mountains that rise sublimely from the valley.

From Keswick there are several roads by which Ullswater may be approached.

1st. By a bridle-road that turns off from the Penrith road at the third milestone, and crosses the Vale of St. John near its foot, then enters the Vale of Wanthwaite, and, after passing through Matterdale, unites at Dockray with

2nd. A good carriage-road that leaves the Penrith road a little beyond the twelfth milestone from Keswick, and skirts the base of a bleak uninteresting mountain called Mell Fell, which the traveller has on his left hand till he reaches the hamlet of Matterdale End, where the road turns sharply to the left to Dockray, before mentioned. From Dockray the traveller will descend upon Gowbarrow Park, and is thus brought at once upon a magnificient view of the higher reaches of the lake. (*See Diagrams, Plate* 5). ARA-FORCE thunders down the ghyll on the left at a small distance from the road. At the foot of the hill, and before proceeding to patterdale, turn in at the gate on the left to LYULPH'S TOWER, where a guide to the Waterfall is always to be had.

3rd. Ullswater may be approached by proceeding direct to Pooley Bridge, at the foot of the lake, where the angler would find much diversion both in the lake and in the neighbouring streams. (*See Diagrams Plate* 3). Pooley Bridge is also favourably situated for visiting Hawes Water, ten miles, and Lowther Castle, four miles; and the town of Penrith, to be hereafter noticed, is only six miles distant.

Besides the approaches to Ullswater, just mentioned, a stout pedestrian might proceed to Patterdale over the northern shoulder of Helvellyn, and visit its summit in his progress, if thought desirable.—In this route, the road to Ambleside must be kept for four miles and three-quarters, whence the road from Wythburn to Threlkeld must be pursued for a short distance to a farm-house called Stainah. The ascent from Stainah, for a considerable distance, is by a steep zig-zag path, on the left of one of the mountain streams falling into St. John's Vale. The road at the top of the first steep turns southward, nearly at right angles, and farther on, at another turn on the left, a few land-marks may be observed, which serve as guides into Patterdale by the Greenside lead mines, in the vale of Glenridding When at the highest part of the foot-road, the Raise, or Styx, a round-topped hill, is on the right; and further to the south, with a considerable dip between them, is another elevation called Whiteside, from whence, by a narrow ridge, the Tourist may proceed to the summit of Helvellyn. The distance, by this road, if Helvellyn be left out, is much less than by any of the former routes, and the views from it are exceedingly impressive. In this excursion strangers would do well to take a guide. See Ascent of Helvellyn from Patterdale.

If Ullswater be approached from Penrith, a mile and a half brings you to the winding Vale of Eamont, and the prospects increase in interest till you reach Patterdale; but the first four miles along Ullswater by this road are comparatively tame.

The following account of Ullswater is from Mr. Wordsworth:—In order to see the lower part of the lake to advantage, it is necessary to go round by Pooley Bridge, and to ride at least three miles along the Westmorland side of the Water, towards Martindale. The views, especially if you ascend from the road into the fields, are magnificent; yet this is only mentioned that the transient visitant may know what exists; for it would be inconvenient to go in search of them. They who take this course of three or four miles *on foot*, should have a boat in readiness at the end of the walk, to carry them across to the Cumberland side of the lake, near Old Church, thence to pursue the road upwards to Patterdale. The Church-yard Yew tree still survives at Old Church, but there are no remains of a Place of Worship, a new Chapel having been erected in a more central situation, which

Chapel was consecrated by the then Bishop of Carlisle, when on his way to crown Queen Elizabeth, he being the only Prelate who would undertake the office. It may be here mentioned, that Bassenthwaite Chapel yet stands in a bay as sequestered as the site of Old Church; such situations having been chosen in disturbed times to elude marauders.

The trunk or body of the Vale of Ullswater need not be further noticed, as its beauties shew themselves: but the curious traveller may wish to know something of its tributary streams.

At Dalemain, about three miles from Penrith, a stream is crossed called the Dacre, or Dacor, which name it bore as early as the time of the Venerable Bede. This stream does not enter the lake, but joins the Eamont a mile below. It rises in the moorish country about Penruddock, and flows down a soft sequestered valley, passing by the ancient mansions of Hutton John and Dacre Castle. The former is pleasantly situated, though of a character somewhat gloomy and monastic, and from some of the fields near Dalemain, Dacre Castle, backed by the jagged summit of Saddleback, with the valley and stream in front, forms a grand picture. There is no other stream that conducts to any glen or valley worthy of being mentioned, till we reach that which leads up to Ara-force, and thence into Matterdale, before spoken of. Matterdale, though a wild and interesting spot, has no peculiar features that would make it worth the stranger's while to go in search of them; but, in Gowbarrow Park the lover of Nature might linger for hours. Here is a powerful brook, which dashes among rocks through a deep glen, hung on every side with a rich and happy intermixture of native wood. Here are beds of luxuriant fern, aged hawthorns, and hollies decked with honeysuckles; and fallow-deer glancing and bounding over the lawns and through the thickets. These are the attractions of the retired views, or constitute a foreground for ever-varying pictures of the majestic lake, forced to take a winding course by bold promontories, and environed by mountains of sublime form, towering above each other. At the outlet of Gowbarrow Park we reach a third stream, which flows through a little recess called Glencoin, where lurks a single house, yet visible from the road. Let the artist or leisurely traveller turn aside to it, for the buildings and objects around them are romantic and picturesque. Having passed under the steeps of Stybarrow Crag, and the remains of its native woods, at Glenridding Bridge, a fourth is crossed, which is contaminated by the operations of the Greenside lead mines in the mountains above.

The opening on the side of Ullswater Vale, down which this stream flows, is adorned with fertile fields, cottages, and natural groves, that agreeably unite with the transverse views of the lake; and the stream, if followed up after the enclosures are left behind, will lead along bold water-breaks and waterfalls to a silent Tarn in the recesses of Helvellyn. But to return to the road in the main Vale of Ullswater.—At the head of the lake (being now in Patterdale) we cross a fifth stream, Grisedale Beck: this would conduct along a woody steep, where may be seen some unusually large ancient hollies, up to the level area of the valley of Grisedale; hence there is a path for foot-travellers, and along which a horse may be led to Grasmere. A sublime combination of mountain forms appears in front while ascending the bed of this valley, and the impression deepens till the path leads almost immediately under the projecting masses of Helvellyn. Having retraced the banks of the stream to Patterdale, and pursued the road up the main Dale, the next considerable stream would, if ascended in the same manner, conduct to Deepdale, the character of which valley may be conjectured from its name. It is terminated by a cove, a craggy and gloomy abyss, with precipitous sides; a faithful receptacle of the snows that are driven into it by the west wind, from the summit of Fairfield. Lastly, having gone along the western side of Brothers-water and passed Hartshop Hall, a stream soon after issues from a cove richly decorated with native wood. This spot is, I believe, never explored by travellers; but, from these sylvan and rocky recesses, whoever looks back on the gleaming surface of Brothers-water, or forward to the precipitous sides and lofty ridges of Dove Crag, &c., will be equally pleased with the grandeur and the wildness of the scenery.

Seven Glens or Valleys have been noticed, which branch off from the Cumberland side of the vale. The opposite side has only two streams of any importance, one of which would lead up from the point where it crosses the Kirkstone road, near the foot of Brothers-water, to the decaying hamlet of Hartshop, remarkable for its cottage architecture, and thence to Hays-water, much frequented by anglers. The other, coming down Martindale, enters Ullswater at Sandwyke, opposite to Gowbarrow Park. No persons but such as come to Patterdale merely to pass through it, should fail to walk as far as Blowick, the only enclosed land which on this side borders the higher part of the lake. The axe has here indiscriminately levelled a rich wood of birches and oaks, that divided this favoured spot into a hundred pictures. It has yet its land-locked bays and rocky promontories; but those beautiful woods are gone, which *perfected* its seclusion; and scenes, that might formerly have been compared to an inexhaustible volume, are now spread before the eye in a single sheet—magnificent indeed, but seemingly perused in a moment! From Blowick a narrow track conducts along the craggy side of Place Fell, richly adorned with juniper, and sprinkled over with birches, to the village of Sandwyke, a few straggling houses, that, with the small estates attached to them, occupy an opening opposite to Lyulph's Tower and Gowbarrow Park. In Martindale, the road loses sight of the lake, and leads over a steep hill, bringing you again into view of Ullswater. Its lowest reach, four miles in length, is before you; and the view terminated by the long ridge of Cross Fell in the distance. Immediately under the eye is a deep-indented bay, with a plot of fertile land, traversed by a small brook, and rendered cheerful by two or three substantial houses of a more ornamented and showy appearance than is usual in those wild spots.

HELVELLYN.

The altitude of Helvellyn is stated, according to the Ordnance Survey, to be 3055 feet above the level of the sea. From the different summits of this mountain comprehensive views are obtained of several of the lakes, and the hills in every direction are thence seen under a more than usually picturesque arrangement.

The ascent is frequently commenced from the inn at Wythburn, on the road from Ambleside to Keswick, the distance from that point being much less than from other places; but the acclivity is too steep for a horse to keep his footing. From Patterdale, however, the ascent, as far as Red Tarn, may, with a little management, be made on horseback, by taking the track up Grisedale, which is approached by a gate on the left, immediately after crossing Grisedale Bridge from the inn. The road leads through the ancient farm-yard of Grasset How, and proceeds, winding up the side of the hill, in the direction of Bleaberry Crag, an offshoot of Striding Edge, which it leaves on the left, and then strikes off by the foot of Red Tarn—

> "—— A cove, a huge recess,
> That keeps, till June, December's snow;
> A lofty precipice in front.
> A silent tarn below,"—

to the stakes where horses are usually tied up while parties proceed to the summit. The road, now, is by ascending Swirrel Edge, a rocky projection of the mountain, crowned by the conical hill called Catchedecam, and a scramble of twenty minutes will place the traveller on the highest point of Helvellyn. Some persons are bold enough, in making the ascent, to traverse the giddy and dangerous height of Striding Edge, a sharp ridge forming the southern boundary of Red Tarn; but this road ought not to be taken by any with weak nerves. The top in many places scarcely affords room to plant the foot, and is beset with awful precipices on either side.*

* Eagles formerly built in the precipitous rock which forms the western barrier of this desolate spot. These birds used to wheel and hover round the head of the solitary angler. It also derives a melancholy interest from the fate of a young man, a stranger, of the name of Gough, who perished, some years ago, by falling down the rocks in his attempt to cross over from Wythburn to Patterdale. His remains were discovered by means of a faithful dog that had lingered here for the space of three months, self-supported, and probably retaining to the last an attachment to the skeleton of its master.

> "This dog had been, through three months' space,
> A dweller in that savage place;
> Yes, proof was plain, that since the day
> On which the traveller thus had died,
> The dog had watch'd about the spot,
> Or by his master's side:
> How nourished here through such long time,
> He knows who gave that love sublime;
> And gave that strength of feeling, great
> Above all human estimate!"

The summit of the mountain is a smooth mossy plain, inclining gently to the west, but terminating abruptly by broken precipices on the east. There are on this mountain two piles of stones (*Men*, as they are called), about a quarter of a mile from each other, and from an angle in the hill between these the best view of the country northward is to be had. Skiddaw, with Saddleback on its right, first claims attention. Nearer the eye, lying in a hollow of the mountain, is Kepple Cove Tarn, bounded on the south by Swirrel Edge and Catchedecam. Further south, between the projecting masses of Swirrel Edge and Striding Edge, lies Red Tarn; and, beyond them, nearly the whole of the middle and lower divisions of Ullswater are seen. On the eastern, or Westmorland, side of Ullswater, are Swarth-fell, Birk-fell, and Place-fell; and over them, looking in a south-easterly direction, may be seen Kidsay Pike, High Street, and Hill Bell; and still further south, and far distant from the eye, the broad top of Ingleborough is visible. Angle Tarn is seen reposing among the hills beyond Patterdale. On the Cumberland side of the lake, Hallsteads, the residence of John Marshall, Esq. is delightfully situated; and, at a greater distance, beyond Penrith, the ridge of Crossfell is stretched out. Looking south, having on the left St. Sunday's Crag, are Scandale fell, Fairfield, and Dolly Wagon Pike: over these summits appear the lakes of Windermere, Coniston, and Esthwaite, with the flat country extending southward to Lancaster. To the right of Dolly Wagon Pike is Seat Sandal, with a patch of Loughrigg fell between them; beyond may be descried the mountains of Coniston, with Black Comb in the distance. Langdale Pikes and Wrynose are seen beyond Steel fell; and, more to the right, over Wythburn head, Scawfell and the Pikes look down in majesty upon their more humble neighbours. Great End and Lingmel Crag project from the vast mass of mountains among which the Pikes on Scawfell stand unrivalled; and nearer the eye are the Borrowdale mountains, Glaramara and Rosthwaite Cam being the most conspicuous. Great Gable rears his head on the right of the Pikes; and more to the north is Kirkfell, over which, on a clear day, the Isle of Man may be seen. Next succeeds the great cluster of mountains extending from Derwent Water to Ennerdale. The first range beyond the heights of

Wythburn are Gate Crag, Maiden Moor, and Cat Bells, all near Derwent Water; and over these are Dale Head and Robinson. On the confines of Buttermere are seen Honister Crag, Fleetwith, Haycocks, High Crag, High Stile, and Red Pike; and still more remote, and north of the Pillar, the Ennerdale Haycocks.

Whitelees Pike, Grassmore, Cawsey Pike, and Grisedale Pike all lie between the above range and the lake of Bassenthwaite, a great part of which lake may be observed from Helvellyn, and beyond Bassenthwaite the distant plains of Cumberland, with the summits of the Scottish mountains. Derwent Water is hid from view.

A fine cool spring of water, called BROWNRIGG WELL, which affords a refreshing draught at all seasons, will be found on the western side of the mountain, about 300 yards from its summit.

PENRITH.

PENRITH is a neat and clean town, situated in a fertile valley, a mile from the confluence of the Eamont and Lowther, with a population of 5385. Market on Tuesday. It is a great thoroughfare, being at the junction of the two great roads from the south to Glasgow and Edinburgh. Penrith and the neighbourhood abound in objects of antiquarian curiosity. In the church-yard there is a monument of great antiquity, called the *Giant's Grave,* consisting of two stone pillars about ten feet high and fifteen feet asunder, and four large semicircular stones, two on each side of the grave, embedded in the earth. The common vulgar report is, that this is the tomb of Sir Ewan or Owen Cæsarius, a gigantic warrior, who reigned in this country in the time of the Saxons. Near this monument there is another antique stone pillar, six feet high, called the *Giant's Thumb.* The Castle is an object of interest, and stands on the west side of the town. It was probably erected by the Neville family in the time of Richard II., as a defence for the inhabitants of the town from their Scottish enemies, and was dismantled in the time of the Commonwealth. The Lancaster and Carlisle Railway skirts the walls of this ancient ruin. The Beacon stands on the summit of a hill on the east side of the town, and is a most conspicuous and interesting object for some distance round Penrith. A curious relic of British antiquity, called ***Arthur's Round Table,*** is to be found about a mile south of the town, on the Westmorland side of the Eamont. It is a circular area twenty-nine yards in diameter, surrounded by a broad ditch and elevated mound, with two approaches cut through the mound opposite to each other. It is supposed to have been an arena for tournaments in the days of chivalry. A few hundred yards to the west of the Round Table is an elevation called ***Mayburgh,*** on which is a circular enclosure one hundred yards in diameter, formed by a broad ridge of rounded stones heaped up to the height of fifteen feet. In the centre of the circle is a rude pillar of stone eleven feet high. This is believed to have been a place of Druidical judicature. There is a more remarkable monument, by some supposed of Druidical times, six miles north-east of Penrith, called ***Long Meg and her Daughters.****[*] It is situated on the summit of a hill near Little Salkeld, and is a circle of three hundred and fifty yards in circumference, formed by seventy-two stones, many of which are ten feet high, with one at the entrance eighteen feet high. —***Brougham Hall,*** the residence of Lord Brougham, stands on a gentle eminence one mile and a half to the south-east of Penrith, and from its situation and beautiful prospects has been styled the "Windsor of the North." The majestic ruins of ***Brougham Castle*** stands on the south of the rivers Eamont and Lowther at their confluence, and are about a mile from Penrith. This castle was anciently the seat of the Veteriponts, and from them descended to the Cliffords and Tuftons: it still belongs to the Earl of Thanet. Camden supposes it to stand on the site of the Roman Station ***Brovoniacum.*** About two miles below Brougham Castle, on the rocky banks of the Eamont, are "two very singular grottos or excavations in a perpendicular rock, by a narrow ledge of which they are alone accessible. One of them is but a small narrow recess, but the other is more capacious, and appears to have had a door and window." It was formerly secured by iron gates, and the marks of iron grating and hinges are still observable upon the rock. These grottos are called the ***Giant's Caves,*** or ***Isis Parlis,*** and in

* See Scenery of the Lakes.

Sandford's MS. Account of Cumberland it is said that Sir Hugh Cæsario lived here, and "was buried in the north side of the church i' th' green field." Five miles from Penrith, near Plumpton are the extensive ruins of *Old Penrith,* formerly a Roman Station, supposed by Camden to be *Petriana,* and by Horsley *Bremetenracum.* INNS, *Crown* and *George.*

LOWTHER CASTLE, the magnificent residence of the Earl of Lonsdale, stands in an extensive park comprising six hundred acres of richly-wooded land, and is five miles south of Penrith. This noble structure is built of pale freestone, and combines the majestic effect of a fortification with the splendour of a regal abode.

"Lowther! in thy majestic Pile are seen
Cathedral pomp and grace, in apt accord
With the baronial castle's sterner mien;
Union significant of God adored,
And charters won and guarded by the sword
Of ancient honour; whence the goodly state
Of polity which wise men venerate,
And will maintain, if God his help afford."

The north and south fronts are of a widely different character, the former presenting the appearance of a castle, and the latter that of a cathedral, with pointed and mullioned windows, delicate pinnacles, niches and cloisters. The scene from this front "accords well with the solemn character of the edifice, being a lawn of emerald green and velvet smoothness, shut in by ornamental trees and shrubs, and by timber of stately growth." The prospect from the north front is more extensive, and that from the great central tower is extremely grand. A high embattled wall surrounds the entrance court, which is approached through an arched gateway. The interior of the Castle is fitted up in a style of splendour corresponding with the richness of the exterior. The grand staircase has an imposing appearance, and the apartments are enriched with a vast quantity of massive plate, and contain several pictures of great value. The building of the Castle was commenced in 1802, from a design by Smirke. Through the liberality of the noble Proprietor it is allowed to be seen by visitors at all seasonable times on application at the lodge.

If, during his tour, the stranger has complained, as he will have had reason to do, of a want of majestic trees, he may be abundantly recompensed for his loss in the far-spreading woods which surround this mansion. Visitants, for the most part, see little of the beauty of these magnificent grounds, being content with the view from the Terrace; but the whole course of the Lowther, from Askham to the bridge under Brougham Hall, presents almost at every step some new feature of river, woodland, and rocky landscape. A portion of this tract has, from its beauty, acquired the name of the Elysian Fields;—but the course of the stream can only be followed by the pedestrian.*

Excursions from Penrith.

To the INN at PATTERDALE.

1½	The Cumberland road runs by Red Hills	1½	1¾	Watermillock	7½
2¼	Dalemain	3¾	1¼	Hallsteads	8¾
2	Junction with the Westmorland road	5¾	2¼	Lyulph's Tower	11
			4	Inn at Patterdale	15

From PENRITH, on the Westmorland side of the Eamont, to POOLEY BRIDGE, and thence on the northern side of Ullswater, to the INN at PATTERDALE.

1¼	Over Eamont Bridge to Arthur's Round Table	1¼	½	Junction with the Cumberland road	6¼
4½	Pooley Bridge	5¾	9¼	Inn at Patterdale	15½

From PENRITH to HAWES WATER.

5	Lowther, or Askham*	5	4	Return by Butterswick	16
7	By Bampton* to Hawes Water ...	12	5	Over Moor Dovack to Powley ...	21
			6	By Dalemain to Penrith	27

To SHAP ABBEY.

5	Askham	5	1	Shap	13
4	Bampton Church	9	11	Penrith	24
3	Shap Abbey	12			

SHAP ABBEY.

Of this once magnificent building, little more than the tower now remains. It was built by Thomas, son of Gospatrick, in the reign of King John, for the Canons of the Præmonstratentian Order, who had been first placed at Preston Patrick, near Kendal. In the neighbourhood of this Abbey is an area upwards of

* The woods about Lowther, and especially near the Mansion, suffered greatly by the hurricane which caused such general devastation of the same kind on the 8th January, 1839.

half a mile in length and twenty or thirty yards broad, formed by huge blocks of granite placed at a distance of ten or twelve yards from each other. This stupendous monument of antiquity is called *Carl Lofts*, and is thought by Pennant to be of Danish origin. Dr. Burn supposes it to have been a Druidical Temple. It is now very much reduced, and can with difficulty be traced, owing to many of the stones having been broken up in clearing the ground for agricultural purposes.

CARLISLE.

Carlisle, the capital of Cumberland, is an ancient city and bishopric. It is situated within eight miles of the Scottish border, and is surrounded by a fertile and open country. Carlisle was a Roman Station, and is within a mile of Hadrian's Wall. In the wars between England and Scotland it was a place of great importance. The town is well built, and many of the streets are very spacious. The Castle is said to have been built in the year 780, and some of the massive and antique buttresses on the north battery are ascribed to William Rufus. Mary Queen of Scots was imprisoned here in 1568, but the rooms she occupied have been recently taken down. The Cathedral is a noble building, and the east window is said to be the largest, as it is certainly the finest, in the island, while the ground is classic, as being the resting-place of the mortal remains of Paley, and the scene of the marriage of the author of Waverley. The new Jail is situated at the southern entrance of the city, contiguous to the County Court-houses, the principal features of which are two magnificent circular towers. A News Room, Reading and Coffee Rooms, have recently been erected from a design by Rickman and Hutchinson, of Birmingham, and are a great ornament to the city. There are extensive cotton works carried on here, and the steam-chimney of Messrs. Dixons' cotton mills is a remarkable object for many miles round. Woollens, linens, and other articles are also manufactured here, and Carlisle is particularly celebrated for its whips and hats. Carlisle is the grand focus of Steam and Railway communication for all parts of the kingdom. It is connected with the Irish Channel by a ship-canal to Bowness, on the Solway, from which port steam-packets are constantly plying to Liverpool, Dublin, Belfast, &c. It is also connected with the West of Cumberland by a railroad to Maryport, Workington, Whitehaven, and the western coast extending to Furness and Ulverston,—with Lancashire, and the South of England by the great trunk *Lancaster and Carlisle* and *London and North Western Railways,*—with Newcastle, Sunderland, and the whole northern coast, by the *Newcastle and Carlisle Railway*, and with the whole of Scotland by the *Caledonian* and other Scottish *Railways.* Population, 21,354. Market on Wednesday and Saturday. INNS, *Bush, Coffee House,* and *Victoria.*

Lanercost Priory, Naworth Castle, and Gillsland Spa, may be conveniently visited from Carlisle by Railway Conveyance.

HEIGHTS OF LAKES ABOVE THE SEA.

	Feet.
Red Tarn (Helvellyn)	2400
Sprinkling Tarn (Borrowdale)	1900
Hawes Water	714
Thirlmere	473
Ullswater	460
Derwent Water	288
Crummock Water	260
Bassenthwaite Water	210
Esthwaite Water	198
Grasmere	196
Wast Water	160
Windermere	115
Coniston Water	106

WATERFALLS.

	Feet.
Scale Force, near Buttermere	160
Colwith Force, five miles from Ambleside	150
Stockgill Force, near Ambleside	152
Lodore Fall, near Keswick	150
Barrow Cascade, near Keswick	122
Dungeon Gill, Langdale	90
Ara Force, Gowbarrow Park	80
Rydal Fall, near Ambleside	70
Birker Force, Eskdale	65
Stanley Gill, Eskdale	62
Nunnery Fall, one mile from Kirkoswald	60
Sour Milk Force, near Buttermere	60
Howk, Caldbeck	50
Skelwith Force	20

HEIGHTS OF THE MOUNTAINS OF THE LAKE DISTRICT.

	Feet.
Scawfell Pike, Cumberland	3166
Scawfell, Cumberland	3100
Helvellyn, Cumberland and Westmorland	3055
Skiddaw, Cumberland	3022
Fairfield, Westmorland	2950
Great Gable, Cumberland	2925
Bowfell, Westmorland	2914
Rydal Head, Westmorland	2910
Pillar, Cumberland	2893
Saddleback	2787
Grassmoor, Cumberland	2756
Red Pike, Cumberland	2750
High Street, Westmorland	2700
Grisedale Pike, Cumberland	2680
Coniston Old Man, Lancashire	2577
Hill Bell, Westmorland	2500
Harrison Stickle, } Langdale Pikes, Westmorland	2400
Pike o' Stickle, } Langdale Pikes, Westmorland	2300
Carrock Fell, Cumberland	2110
High Pike, Caldbeck Fells, Cumberland	2101
Causey Pike, Cumberlaud	2030
Black Comb, Cumberland	1919
Lord's Seat, Cumberland	1728
Wansfell, Westmorland	1590
Whinfell Beacon, near Kendal, Westmorland	1500
Cat Bell, Cumberland	1448
Latrigg, Cumberland	1160
Dent Hill, Cumberland	1110
Loughrigg Fell, Westmorland	1108
Benson Knott, near Kendal, Westmorland	1098
Penrith Beacon, Cumberland	1020
Mell Fell, Cumberland	1000
Kendal Fell, Westmorland	648
Scilly Bank, near Whitehaven, Cumberland	500

MOUNTAIN PASSES.

Sty Head, Cumberland	1250
Haws between Buttermere dale and Newlands, Cumberland	1160
Haws between Buttermere and Borrowdale, Cumberland	1100
Dunmail Raise, Cumberland and Westmorland	720

DESCRIPTION

OF

The Scenery of the Lakes.

SCENERY OF THE LAKES.

SECTION FIRST.

VIEW OF THE COUNTRY AS FORMED BY NATURE.

At Lucerne, in Switzerland, is shewn a Model of the Alpine country which encompasses the Lake of the Four Cantons. The spectator ascends a little platform, and sees mountains, lakes, glaciers, rivers, woods, waterfalls, and valleys, with their cottages, and every other object contained in them, lying at his feet; all things being represented in their appropriate colours. It may be easily conceived that this exhibition affords an exquisite delight to the imagination, tempting it to wander at will from valley to valley, from mountain to mountain, through the deepest recesses of the Alps. But it supplies also a more substantial pleasure; for the sublime and beautiful region, with all its hidden treasures, and their bearings and relations to each other, is thereby comprehended and understood at once.

Something of this kind, without touching upon minute details and individualities which would only confuse and embarrass, will here be attempted, in respect to the Lakes of the North of England, and the vales and mountains enclosing and surrounding them. The delineation, if tolerably executed, will, in some instances, communicate to the traveller, who has already seen the objects, new information; and will assist in giving to his recollections a more orderly arrangement than his own opportunities of observing may have permitted him to make; while it will be still more useful to the future traveller, by directing his attention at once to distinctions in things which, without such previous aid, a length of time only could enable him to discover. It his hoped, also, that this Essay may become generally serviceable, by leading to habits of more exact and considerate observation than, as far as the writer knows, have hitherto been applied to local scenery.

To begin, then, with the main outlines of the country:—I know not how to give the reader a distinct image of these more readily, than by requesting him to place himself with me, in imagination, upon some given point: let it be the top of either of the mountains, Great Gable, or Scawfell: or, rather, let us suppose our station to be a cloud hanging midway between these two mountains, at not more than half a mile's distance from the summit of each, and not many yards above their highest elevation; we shall than see stretched at our feet a number of valleys, not fewer than eight, diverging from the point on which we are supposed to stand, like spokes from the nave of a wheel. First, we note, lying to the south-east, the vale of Langdale,* which will conduct the eye to the long lake of Windermere, stretching near to the sea; or rather to the sands of the vast bay of Morecambe, serving here for the rim of this imaginary wheel: let us trace it in a direction from the south-east towards the south, and we shall next fix our eyes upon the vale of Coniston, running up likewise from the sea, but not (as all the other valleys do) to the nave of the wheel, and therefore it may be not inaptly represented as a broken spoke sticking in the rim. Looking forth again, with an inclination towards the west, we see immediately at our feet the vale of Duddon, in which is no lake, but a copious stream winding among fields, rocks, and mountains, and terminating its course in the sands of Duddon. The fourth vale, next to be observed, viz, that of the Esk, is of the same general character as the last, yet beautifully discriminated from it by peculiar features. Its stream passes under the woody steep upon which stands Muncaster Castle, the ancient seat of the Penningtons, and after forming a short and narrow æstuary enters the sea below the small town of Ravenglass. Next, almost due west, look down into and along the deep valley of Wastdale, with its little chapel and half a dozen neat dwellings scattered upon a plain of meadow and corn-ground intersected with stone walls apparently innumerable, like a large piece of lawless patchwork, or an array of mathematical figures, such as in the ancient schools of geometry might have been sportively and fantastically traced out upon sand. Beyond this little fertile plain lies, within a bed of steep mountains, the long, narrow, stern, and desolate lake of Wastdale; and, beyond this, a dusky tract of level ground conducts the eye to the Irish Sea. The stream that issues from Wastwater is named the Irt, and falls into the æstuary of the river Esk. Next comes in view Ennerdale, with its lake of bold and somewhat savage shores. Its stream, the Ehen or Enna, flowing through a soft and fertile country, passes the town of Egremont and the ruins of the castle,—then, seeming, like the other rivers, to break through the barrier of sand thrown up by the winds on this tempestuous coast, enters the Irish Sea. The vale of Buttermere, with the lake and village of that name, and

* Anciently spelt Langden, and so called by the old inhabitants to this day—*dean*, from which the latter part of the word is derived, being in many parts of England a name for a valley.

Crummock-water, beyond, next present themselves. We will follow the main stream, the Cocker, through the fertile and beautiful vale of Lorton, till it is lost in the Derwent, below the noble ruins of Cockermouth Castle. Lastly, Borrowdale, of which the vale of Keswick is only a continuation, stretching due north, brings us to a point nearly opposite to the vale of Windermere, with which we began. From this it will appear, that the image of a wheel, thus far exact, is a little more than one half complete: but the deficiency on the eastern side may be supplied by the vales of Wythburn, Ullswater, Haweswater, and the vale of Grasmere and Rydal; none of these, however, run up the central point between Great Gable and Scawfell.

From this, hitherto our central point, let us take flight of not more than four or five miles eastward to the ridge of Helvellyn, and we shall look down upon Wythburn and St. John's Vale, which are a branch of the vale of Keswick; upon Ullswater, stretching due east; and not far beyond to the south-east (though from this point not visible) lie the vale and lake of Haweswater; and lastly, the vale of Grasmere, Rydal, and Ambleside, brings us back to Windermere, thus completing, though on the eastern side in a somewhat irregular manner, the representative figure of the wheel.

Such, concisely given, is the general topographical view of the country of the Lakes in the north of England; and it may be observed that, from the circumference to the centre, that is, from the sea, or plain country, to the mountain stations specified, there is—in the several ridges that enclose these vales and divide them from each other, I mean in the forms and surfaces, first of the swelling grounds, next of the hills and rocks, and lastly of the mountains—an ascent of almost regular gradation, from elegance and richness, to their highest point of grandeur and sublimity. It follows, therefore, from this, first, that these rocks, hills, and mountains, must present themselves to view in stages rising above each other, the mountains clustering together towards the central point; and next, that an observer familiar with the several vales must, from their various positions in relation to the sun, have had before his eyes every possible embellishment of beauty, dignity, and splendour, which light and shadow can bestow upon objects so diversified. For example, in the vale of Windermere, if the spectator looks for gentle and lovely scenes, his eye is turned towards the south; if for the grand, towards the north: in the vale of Keswick, which (as hath been said) lies almost due north of this, it is directly the reverse. Henee, when the sun is setting in summer far to the north-west, it is seen by the spectator from the shores or breast of Windermere, resting among the summits of the loftiest mountains, some of which will perhaps be half or wholly hidden by clouds, or by the blaze of light which the orb diffuses around it; and the surface of the lake will reflect before the eye corresponding colours through every variety of beauty, and through all degrees of splendour. In the vale of Keswick, at the same period, the sun sets over the humbler regions of the landscape, and showers down upon *them* the radiance which at once veils and glorifies,—sending forth, meanwhile, broad streams of rosy, crimson, purple, or golden light, towards the grand mountains in the south and south-east, which, thus illuminated, with all their projections and cavities, and with the intermixture of solemn shadows, are seen distinctly through a cool and clear atmosphere. Of course, there is as marked a difference between the *noontide* appearance of these two opposite vales. The bedimming haze that overspreads the south, and the clear atmosphere and determined shadows of the clouds in the north, at the same time of the day, are each seen in the several vales, with a contrast as striking. The reader will easily conceive in what degree the intermediate vales partake of a kindred variety.

I do not indeed know any tract of country in which, within so narrow a compass, may be found an equal variety in the influences of light and shadow upon the sublime or beautiful features of landscape: and it is owing to the combined circumstances to which the reader's attention has been directed. From a point between Great Gable and Scawfell, a shepherd would not require more than an hour to descend into any one of eight of the principal vales by which he would be surrounded; and all the others lie (with the exception of Haweswater) at but a small distance. But, though clustered together, every valley has its distinct and separate character: in some instances, as if they had been formed in studied contrast to each other, and in others with the united pleasing differences and resemblances of a sisterly rivalship. This concentration of interest gives to the country a decided superiority over the most attractive districts of Scotland and Wales, especially for the pedestrian traveller. In Scotland and Wales are found, undoubtedly, individual scenes, which, in their several kinds, cannot be excelled. But, in Scotland, particularly, what long tracts of desolate country intervene! so that the traveller, when he reaches a spot deservedly of great celebrity, would find it difficult to determine how much of his pleasure is owing to excellence inherent in the landscape itself; and how much to an instantaneous recovery from an oppression left upon his spirits by the barrenness and desolation through which he has passed.

But to proceed with our survey; and, first, of the MOUNTAINS. Their *forms* are endlessly diversified, sweeping easily or boldly in simple majesty, abrupt and precipitous, or soft and elegant. In magnitude and grandeur they are individually inferior to the most celebrated of those in some other parts of this island; but, in the combinations which they make, towering above each other, or lifting themselves in ridges like the waves of a tumultuous sea, and in the beauty and variety of their surfaces and colours, they are surpassed by none.

The general *surface* of the mountains is turf, rendered rich and green by the moisture of the climate. Sometimes the turf, as in the neighbourhood of Newlands, is little broken, the whole covering being soft and downy pasturage. In other places, rocks predominate; the soil is laid bare by torrents and burstings of water from the sides of the mountains in heavy rains; and not unfrequently their perpendicular sides are seamed by ravines (formed also by rains and torrents), which, meeting in angular points, entrench and scar the surface with numerous figures like the letters W and Y.

In the ridge that divides Eskdale from Wastdale, granite is found; but the mountains are for the most part composed of the stone by mineralogists termed schist, which, as you approach the plain country, gives place to limestone and freestone; but schist being the substance of the mountains, the predominant *colour* of their *rocky* parts is bluish, or hoary grey—the general tint of the lichens with which the bare stone is encrusted. With this blue or grey colour is frequently intermixed a red tinge, proceeding from the iron that interveins the stone and impregnates the soil. The iron is the principle of decomposition in these rocks; and hence, when they become pulverized, the elementary particles crumbling down, overspread in many places the steep and almost precipitous sides of the mountains with an intermixture of colours, like the compound hues of the dove's neck. When in the heat of advancing summer, the fresh green tint of the herbage has somewhat faded, it is again revived by the appearance of the fern profusely spread over the same ground; and, upon this plant, more than upon any thing else, do the changes which the seasons make in the colouring of the mountains depend. About the first week in October, the rich green, which prevailed through the whole summer, is usually passed away. The brilliant and various colours of the fern are then in harmony with the autumnal woods: bright yellow or lemon colour, at the base of the mountains, melting gradually, through orange, to a dark russet brown towards the summits, where the plant, being more exposed to the weather, is in a more advanced state of decay. Neither heath nor furz are *generally* found upon the *sides* of these mountains, though in many places they are adorned by those plants, so beautiful when in flower. We may add, that the mountains are of height sufficient to have the surface towards the summit softened by distance, and to imbibe the finest aërial hues. In common also with other mountains, their apparent forms and colours are perpetually changed by the clouds and vapours which float round them: the effect indeed of mist or haze, in a country of this character, is like that of magic. I have seen six or seven ridges rising above each other, all created in a moment by the vapours upon the side of a mountain, which in its ordinary appearance showed not a projecting point to furnish even a hint for such an operation.

I will take this opportunity of observing, that they who have studied the appearance of nature feel that the superiority, in point of visual interest, of mountainous over other countries—is more strikingly displayed in winter than in summer. This, as must be obvious, is partly owing to the *forms* of the mountains, which, of course, are not affected by the seasons; but also, in no small degree, to the greater variety that exists in their winter than their summer *colouring*. This variety is such, and so harmoniously preserved, that it leaves little cause of regret when the splendour of autumn is passed away. The oak coppices, upon the sides of the mountains, retain russet leaves; the birch stands conspicuous with its silver stem and puce-coloured twigs; the hollies, with green leaves and scarlet berries, have come forth to view from the deciduous trees, whose summer-foliage had concealed them: the ivy is now plentifully apparent upon the stems and boughs of the trees, and upon the steep rocks. In place of the deep summer-green of the herbage and fern, many rich colours play into each other over the surface of the mountains; turf (the tints of which are interchangeably tawny-green. olive, and brown), beds of withered fern, and grey rocks, being harmoniously blended together. The mosses and lichens are never so fresh and flourishing as in winter, if it be not a season of frost; and their minute beauties prodigally adorn the foreground. Wherever we turn, we find these productions of nature, to which winter is rather favourable than unkindly, scattered over the walls, banks of earth, rocks, and stones, and upon the trunks of trees, with the intermixture of several species of small fern, now green and fresh; and, to the observing passenger, their forms and colours are a source of inexhaustible admiration. Add to this the hoar-frost and snow, with all the varieties they create, and which volumes would not be sufficient to describe. I will content myself with one instance of the colouring produced by snow, which may not be uninteresting to painters. It is extracted from the memorandum-book of a friend; and for its accuracy I can speak, having been an eye-witness of the appearance. "I observed," says he, "the beautiful effect of the drifted snow upon the mountains, and the perfect *tone* of colour. From the top of the mountains downwards, a rich olive was produced by the powdery snow and the grass, which olive was warmed with a little brown, and in this way harmoniously combined, by insensible gradations, with the white. The drifting took away the monotony of snow; and the whole vale of Grasmere, seen from the terrace walk in Easedale, was as varied, perhaps more so, than even in the pomp of autumn. In the distance was Loughrigg Fell, the basin-wall of the lake: this, from the summit downward, was a rich orange-olive; then the lake of a bright olive-green, nearly the same tint as the snow-powdered mountain tops and high slopes in Easedale; and, lastly, the church, with its firs, forming the centre of the view. Next to the church came nine distinguishable hills, six of them with woody sides turned towards us, all of them oak copses with their bright red leaves and snow-powdered twigs; these hills—so variously situated in relation to each other, and to the view in general, so variously powdered, some only enough to give the herbage a rich brown tint, one intensely white and lighting up all the others—were yet so placed, as in the most inobtrusive manner to harmonise by contrast with a perfect naked, snowless, bleak summit in the far distance."

Having spoken of the forms, surface, and colour of the mountains, let us descend into the VALES. Though these have been represented under the general image of the spokes of a wheel, they are, for the most part, winding; the windings of many being abrupt and intricate. And it may be observed, that, in one circumstance, the general shape of them all has been determined by that primitive conformation through which so many became receptacles of lakes. For they are not formed, as are most of the celebrated Welch valleys, by an approximation of the sloping bases of the opposite mountains towards each other, leaving little more between than a channel for the passage of a hasty river; but the bottom of these valleys is mostly a spacious and gently declining area, apparently level as the floor of a temple, or the surface of a lake, and broken in many cases by rocks and hills, which rise up like islands from the plain. In such of the valleys as may make windings, these level areas open upon the traveller in succession, divided from each other sometimes by a mutual approximation of the hills, leaving only passage for a river, sometimes by correspondent windings, without such approximation; and sometimes by a bold advance of one mountain to that which is opposite it. It may here be observed with propriety that the several rocks and hills, which have been described as rising up like islands from the level area of the vale, have regulated the choice of the inhabitants in the situation of their dwellings. Where none of these are found, and the inclination of the ground is not sufficiently rapid easily to carry off the waters (as in the higher part of Langdale, for instance) the houses are not sprinkled over the middle of the vales, but confined to their sides, being placed merely so far up the mountain as to be protected from the floods. But where these rocks and hills have been scattered over the plain of the vale (as in Grasmere, Donnerdale, Eskdale, &c.) the beauty they give to the scene is much heightened by a single cottage, or cluster of cottages, that will be almost always found under them, or upon their sides; dryness and shelter having tempted the dalesmen to fix their habitations there.

I shall now speak of the LAKES of this country. The form of the lake is most perfect when, like Derwent-water and some of the smaller lakes, it least resembles that of a river;—I mean, when being looked at from any given point where the whole may be seen at once, the width of it bears such proportion to the length, that, however the outline may be diversified by far-receding bays, it never assumes the shape of a river, and is contemplated with that placid and quiet feeling which belongs peculiarly to the lake—as a body of still water under the influence of no current; reflecting therefore the clouds, the light, and all the imagery of the sky and surrounding hills; expressing also and making visible the changes of the atmosphere and motions of the lightest breeze, and subject to agitation only from the winds,

> ————————The visible scene
> Would enter unawares into his mind
> With all its solemn imagery, its rocks,
> Its woods, and that uncertain heaven received
> Into the bosom of the *steady* lake!

It must be noticed, as a favourable characteristic of the lakes of this country, that, though several of the largest, such as Windermere, Ullswater, and Haweswater, do, when the whole length of them is commanded from an elevated point, lose somewhat of the peculiar form of the lake, and assume the resemblance of a magnificent river; yet, as their shape is winding (particularly that of Ullswater and Haweswater), when the view of the whole is obstructed by those barriers which determine the windings, and the spectator is confined to one reach, the appropriate feeling is revived; and one lake may thus in succession present to the eye the essential characteristic of many. But though the forms of the large lakes have this advantage, it is nevertheless favourable to the beauty of the country that the largest of them are comparatively small: and that the same vale generally furnishes a succession of lakes, instead of being filled with one. The vales in North Wales, as hath being observed, are not formed for the reception of lakes; those of Switzerland, Scotland, and this part of the North of England, *are* so formed; but in Switzerland and Scotland, the proportion of diffused water is often too great, as at the lake of Geneva for instance, and in most of the Scotch lakes. No doubt it sounds magnificent and flatters the imagination, to hear, at a distance, of expanses of water so many leagues in length and miles in width; and such ample room may be delightful to the fresh-water sailor, scudding with a lively breeze amid the rapidly-shifting scenery. But, who ever travelled along the banks of Loch-Lomond, variegated as the lower part is by islands, without feeling that a speedier termination of the long vista of blank water would be acceptable; and without wishing for an interposition of green meadows, trees, and cottages, and a sparkling stream to run by his side? In fact, a notion of grandeur, as connected with magnitude, has seduced persons of taste into general mistake upon this subject. It is much more desirable, for the purpose of pleasure, that lakes should be numerous, and small or middle-sized, than large, not only for communication by walks and rides, but for variety, and for recurrence of similar appearances. To illustrate this by one instance:—how pleasing is it to have a ready and frequent opportunity of watching, at the outlet of a lake, the stream pushing its way among the rocks in lively contrast with the stillness from which it has escaped; and how amusing to compare its noisy and turbulent motions with the gentle playfulness of the breezes that may be starting or wandering here and there over the faintly-rippled surface of the broad water! I may add as a general remark, that, in lakes of great width, the shores cannot be distinctly seen at the same time, and therefore contribute little to mutual illustration and ornament; and, if the opposite shores are out of sight of each other, like those of the American and Asiatic lakes, then, unfortunately, the traveller is reminded of a nobler object; he has the blankness of a sea-prospect without the grandeur and accompanying sense of power.

As the comparatively small size of the lakes in the North of England is favourable to the production of varigated landscapes, their *boundary-line* also is for the most part gracefully or boldly indented. That uniformity which prevails in the primitive frame of the lower grounds among all chains or clusters of mountains where large bodies of still water are bedded, is broken by the *secondary* agents of nature, ever at work to supply the deficiencies of the mould in which things were originally cast. Using the word *deficiencies*, I do not speak with reference to those stronger emotions which a region of mountains is peculiarly fitted to excite. The bases of these huge barriers may run for a long space in straight lines, and these parallel to each other; the opposite sides of a profound vale may ascend as exact counterparts, or in mutual reflection, like the billows of a troubled sea; and the impression be, from its very simplicity, more awful and sublime. Sublimity is the result of Nature's first great dealings with the superficies of the earth; but the general tendency of her subsequent operations is towards the production of beauty, by a multiplicity of symmetrical parts uniting in a consistent whole This is everywhere exemplified along the margins of these lakes. Masses of rock, that have been precipitated from the heights into the area of waters, lie in some places like stranded ships; or have acquired the compact structure of jutting piers; or project in little peninsulas crested with native wood. The smallest rivulet—one whose silent influx is scarcely noticeable in a season of dry weather—so faint is the dimple made by it on the surface of the smooth lake—will be found to have been not useless in shaping, by its deposits of gravel and soil in time of flood, a curve that would not otherwise have existed. But the more powerful brooks, encroaching upon the level of the lake, have, in course of time, given birth to ample promontories of sweeping outline that contrast boldly with the longitudinal base of the steeps on the opposite shore; while their flat or gently-sloping surfaces never fail to introduce, into the midst of desolation and barrenness, the elements of fertility, even where the habitations of men may not have been raised. These alluvial promontories, however, threaten, in some places, to bisect the waters which they have long adorned; and, in course of ages, they will cause some of the lakes to dwindle into numerous and insignificant pools, which, in their turn, will be finally filled up. But, checking these intrusive calculations, let us rather be content with appearances as they are, and pursue in imagination the meandering shores; whether rugged steeps, admitting of no cultivation, descend into the water, or gently-sloping lawns and woods, or flat and fertile meadows stretch between the margin of the lake and the mountains. Among minuter recommendations will be noticed, especially along bays exposed to the setting-in of strong winds, the curved rim of fine blue gravel, thrown up in course of time by the waves, half of it perhaps gleaming from under the water, and the corresponding half of a lighter hue; and in other parts bordering the lake, groves, if I may so call them, of reeds and bulrushes; or plots of water-lilies lifting up their large target-shaped leaves to the breeze, while the white flower is heaving upon the wave.

To these may naturally be added the Birds that enliven the waters. Wild ducks in spring-time hatch their young in the islands, and upon reedy shores;—the sand-piper, flitting along the stoney margins, by its restless note attracts the eye to motions as restless:—upon some jutting rock, or at the edge of a smooth meadow, the stately heron my be descried with folded wings, that might seem to have caught their delicate hue from the blue waters, by the side of which she watches for her sustenance. In winter, the lakes are sometimes resorted to by wild swans; and in that season habitually by widgeons, goldings, and other aquatic fowl of the smaller species. Let me be allowed the aid of verse to describe the evolutions which these visitants sometimes perform on a fine day towards the close of winter.

Mark how the feather'd tenants of the flood,
With grace of motion that might scarcely seem
Inferior to angelical, prolong
Their curious pastime! shaping in mid-air
(And sometimes with ambitious wing that soars
High as the level of the mountain tops,)
A circuit ampler than the lake beneath,
Their own domain;—but ever, while intent
On tracing and retracing that large round,
Their jubilant activity evolves
Hundreds of curves and circlets, to and fro,
Upward and downward, progress intricate
Yet perplex'd, as if one spirit swayed
Their indefatigable flight.—'Tis done—
Ten times, or more, I fancied it had ceased;
But lo! the vanish'd company again
Ascending;—they approach—I hear their wings
Faint, faint, at first, and then an eager sound
Past in a moment—and as faint again!
They tempt the sun to sport amid their plumes:
They tempt the water or the gleaming ice,
To show them a fair image;—'tis themselves,
Their own fair forms, upon the glimmering plain,
Painted more soft and fair as they descend
Almost to touch;—then up again aloft,
Up with a sally and a flash of speed,
As if they scorned both resting-place and rest!

The ISLANDS, dispersed among these lakes, are neither so numerous nor so beautiful as might be expected from the account that has been given of the manner in which the level areas of the vales are so frequently diversified by rocks, hills and hillocks scattered over them; nor are they ornamented (as are several of the lakes in Scotland and Ireland) by the remains of castles or other places of defence; nor with the still more interesting ruins of religious edifices. Every one must regret that scarcely a vestige is left of the Oratory, consecrated to the Virgin, which stood upon Chapel-Holm, in Windermere, and that the Chantry has disappeared, where mass used to be sung, upon St. Herbert's Island, Derwentwater. The islands of the last-mentioned lake are neither fortunately placed nor of pleasing shape; but if the wood upon them were managed with more taste, they might become interesting features in the landscape. There is a beautiful cluster on Windermere; a pair pleasingly contrasted upon Rydal: nor must the solitary green island of Grasmere be forgotten.

In the bosom of each of the lakes of Ennerdale and Devockwater is a single rock, which, owing to its neighbourhood to the sea, is

"The haunt of cormorants and sea-mew's clang."

a music well suited to the stern and wild character of the several scenes. It may be worth while here to mention (not as an object of beauty, but of curiosity), that there occasionally appears above the surface of Derwentwater, and always in the same place, a considerable tract of spongy ground covered with aquatic plants, which is called the Floating, but with more propriety might be named the Buoyant, Island; and, on one of the pools near the lake of Esthwaite, may sometimes be seen a mossy Islet, with trees upon it, shifting about before the wind, a *lusus naturæ* frequent on the great rivers of America, and not unknown in other parts of the world.

——— "fas habeas invisere Tiburis arva,
Albuneæque lacum, atque umbras terrasque natantes."*

This part of the subject may be concluded with observing—that from the multitude of brooks and torrents that fall into these lakes, and of internal springs by which they are fed, and which circulate through them like veins, they are truly living lakes, "*vivi lacus*;" and are thus discriminated from the stagnant and sullen pools frequent among mountains that have been formed by volcanoes, and from the shallow meres found in flat and fenny countries. The water is also of crystalline purity; so that, if it were not for the reflections of the incumbent mountains by which it is darkened, a delusion might be felt, by a person resting quietly in a boat on the bosom of Windermere or Derwentwater, similar to that which Carver so beautifully describes when he was floating alone in the middle of lake Erie or Ontario, and could almost have imagined that his boat was suspended in an element as pure as air, or, rather, that the air and water were one.

Having spoken of Lakes, I must not omit to mention, as a kindred feature of this country, those bodies of still water called TARNS. In the economy of nature these are useful, as auxiliars to Lakes; for if the whole quantity of water which falls upon the mountains in time of storm were poured down upon the plains without the intervention, in some quarters, of such receptacles, the habitable grounds would be much more subject than they are to inundation. But, as some of the collateral brooks spend their fury, finding a free course toward, and also down the channel of the main stream of the vale, before those that have to pass through the higher tarns and lakes have filled their several basins, a gradual distribution is effected; and the waters thus reserved, instead of uniting to spread ravage and deformity with those

* See the Catillus and Salia of Landor.

which meet with no such detention, contribute to support, for a length of time, the vigour of many streams without a fresh fall of rain. Tarns are found in some of the vales, and are numerous upon the mountains. A Tarn, in a *Vale*, implies, for the most part, that the bed of the vale is not happily formed; that the water of the brooks can neither wholly escape, nor diffuse itself over a large area. Accordingly, in such situations, Tarns are often surrounded by an unsightly tract of boggy ground; but this is not always the case, and in the cultivated parts of the country, when the shores of the Tarn are determined, it differs only from the Lake in being smaller, and in belonging mostly to a smaller valley, or circular recess. Of this class of miniature lakes, Loughrigg Tarn, near Grasmere, is the most beautiful example. It has a margin of green firm meadows, of rocks, and rocky woods, a few reeds here, a little company of water-lillies there, with beds of gravel or stone beyond; a tiny stream issuing neither briskly nor sluggishly out of it; but its feeding rills, from the shortness of their course, so small as to be scarcely visible. Five or six cottages are reflected in its peaceful bosom; rocky and barren steeps rise up above the hanging enclosures; and the solemn pikes of Langdale overlook, from a distance, the low cultivated ridge of land that forms the northern boundary of this small, quiet, and fertile domain. The *Mountain* Tarns can only be recommended to the notice of the inquisitive traveller who has time to spare. They are difficult of access and naked; yet some of them are, in their permanent forms, very grand; and there are accidents of things which would make the meanest of them interesting. At all events, one of these pools is an acceptable sight to the mountain wanderer; not merely as an incident that diversifies the prospect, but as forming in his mind a centre or conspicuous point, to which objects, otherwise disconnected or insubordinated, may be referred. Some few have a varied outline, with bold heath-clad promontories; and, as they mostly lie at the foot of a steep precipice, the water, where the sun is not shining upon it, appears black and sullen: and, round the margin, huge stones and masses of rock are scattered; some defying conjecture as to the means by which they came thither; and others obviously fallen from on high—the contribution of ages! A not unpleasing sadness is induced by this perplexity, and these images of decay; while the prospect of a body of pure water unattended with groves and other cheerful rural images by which fresh water is usually accompanied, and unable to give furtherance to the meagre vegetation around it—excites a sense of some repulsive power strongly put forth, and thus deepens the melancholy natural to such scenes. Nor is the feeling of solitude often more forcibly or more solemnly impressed than by the side of one of these mountain pools: though desolate and forbidding, it seems a distinct place to repair to; yet where the visitants must be rare, and there can be no disturbance. Water-fowl flock hither; and the lonely angler may sometimes here be seen; but the imagination, not content with this scanty allowance of society, is tempted to attribute a voluntary power to every change which takes place in such a spot, whether it be the breeze that wanders over the surface of the water, or the splendid lights of evening resting upon it in the midst of awful precipices.

> "There, sometimes, does a leaping fish
> Send through the tarn a lonely cheer;
> The crags repeat the raven's croak
> In symphony austere;
> Thither the rainbow comes,—the cloud,—
> And mists that spread the flying shroud,
> And sunbeams, and the sounding blast."

It will be observed that this country is bounded on the south and east by the sea, which combines beautifully, from many elevated points, with the inland scenery; and, from the bay of Morecambe, the sloping shores and background of distant mountains are seen, composing pictures equally distinguished for amenity and grandeur. But the æstuaries on this coast are in a great measure bare at low water,* and there is no instance of the sea running far up among the mountains, and mingling with the lakes, which are such in the strict and usual sense of the word, being of fresh water. Nor have the streams, from the shortness of their course, time to acquire that body of water necessary to confer upon them such majesty. In fact, the most considerable, while they continue in the mountain and lake country, are rather large brooks than rivers. The water is perfectly pellucid, through which in many places are seen, to a great depth, their beds of rock, or of blue gravel, which give to the water itself an exquisitely cerulean colour; this is particularly striking in the rivers Derwent and Duddon, which may be compared, such and so various are their beauties, to any two rivers of equal length of course in any country. The number of the torrents and smaller brooks is infinite, with their water-falls and water-breaks; and they need not here be described. I will only observe that, as many, even of the smallest rills, have either found, or made for themselves, recesses in the sides of the mountains or in the vales,

* In fact there is not an instance of a harbour on the Cumberland side of the Solway Frith that is not dry at low water: that of Ravenglass, at the mouth of the Esk, as a natural harbour, is much the best. The sea appears to have been retiring slowly for ages from this coast. From Whitehaven to St. Bees extends a tract of level ground, about five miles in length, which formerly must have been under salt water, so as to have made an island of the high ground that stretches between it and the sea.

they have tempted the primitive inhabitants to settle near them for shelter; and hence, cottages so placed, by seeming to withdraw from the eye, are more endeared to the feelings.

The Woods consist chiefly of oak, ash, and birch, and here and there wych-elm, with underwood of hazle, the white and black thorn, and hollies; in moist places alders and willows abound; and yews among the rocks. Formerly the whole country must have been covered with wood to a great height up the mountains; where native Scotch firs* must have grown in great profusion, as they do in the northern part of Scotland to this day. But not one of these old inhabitants has existed, perhaps, for some hundreds of years; the beautiful traces, however, of the universal sylvan† appearance the country formerly had, yet survive in the native coppice-woods that have been protected by inclosures, and also in the forest-trees and hollies, which, though disappearing fast, are yet scattered both over the inclosed and uninclosed parts of the mountains. The same is expressed by the beauty and intricacy with which the fields and coppice-woods are often intermingled; the plough of the first settlers having followed naturally the veins of richer, dryer, or less stony soil; and thus it has shaped out an intermixture of wood and lawn, with a grace and wildness which it would have been impossible for the hand of studied art to produce. Other trees have been introdued within these last fifty years, such as beeches, larches, limes, &c., and plantations of firs, seldom with advantage, and often with great injury to the appearance of the country; but the sycamore (which I believe was brought into this island from Germany, not more than two hundred years ago) has long been the favourite of the cottagers; and, with the fir, has been chosen to screen their dwellings; and is sometimes found in the fields whither the winds or the waters may have carried its seeds.

The want most felt, however, is that of timber trees. There are few *magnificent* ones to be found near any of the lakes; and unless greater care be taken, there will, in a short time, scarcely be left an ancient oak that would repay the cost of felling. The neighbourhood of Rydal, notwithstanding the havoc which has been made, is yet nobly distinguished. In the woods of Lowther, also, is found an almost matchless store of ancient trees, and the majesty and wildness of the native forest.

Among the smaller vegetable ornaments must be reckoned the bilberry, a ground plant, never so beautiful as in early spring,

* This species of fir is in character much superior to the American, which has usurped its place. Where the fir is planted for ornament, let it be by all means of the aboriginal species, which can only be procured from the Scotch nurseries.

† A squirrel (so I have heard the old people of Wythburn say) might have gone from their chapel to Keswick without alighting on the ground.

when it is seen under bare or budding trees, that imperfectly intercept the sunshine, covering the rocky knolls with a pure mantle of fresh verdure, more lively than the herbage of the open fields:—the broom that spreads luxuriantly along rough pastures, and in the month of June interveins the steep copses with its golden blossoms; and the juniper, a rich evergreen, that thrives, in spite of cattle, upon the uninclosed parts of the mountains; the Dutch myrtle diffuses fragrance in moist places; and there is an endless variety of brilliant flowers in the fields and meadows, which, if the agriculture of the country were more carefully attended to, would disappear. Nor can I omit again to notice the lichens and mosses: their profusion, beauty, and variety exceed those of any other country I have seen.

It may now be proper to say a few words respecting CLIMATE and "skiey influences," in which this region, as far as the character of its landscapes is affected by them, may, upon the whole, be considered fortunate. The country is, indeed, subject to much bad weather, and it has been ascertained that twice as much rain falls here as in many parts of the island; but the number of black drizzling days, that blot out the face of things, is by no means *proportionally* great. Nor is a continuance of thick, flagging, damp air so common as in the west of England and Ireland. The rain here comes down heartily, and is frequently succeeded by clear, bright weather, when every brook is vocal, and every torrent sonorous; brooks and torrents which are never muddy, even in the heaviest floods, except, after a draught, they happen to be defiled for a short time by waters that have swept along dusty roads, or have broken out into ploughed fields. Days of unsettled weather, with partial showers, are frequent; but the showers darkening, or brightening, as they fly from hill to hill, are not less grateful to the eye than finely interwoven passages of gay and sad music are touching to the ear. Vapours exhaling from the lakes and meadows after sun-rise, in a hot season, or in moist weather, brooding upon the heights, or descending towards the valleys with inaudible motion, give a visionary character to every thing around them; and are in themselves so beautiful as to dispose us to enter into the feelings of those simple nations (such as the Laplanders of this day) by whom they are taken for guardian deities of the mountains; or to sympathise with others who have fancied these delicate apparitions to be the spirits of their departed ancestors. Akin to these are fleecy clouds resting upon the hill tops: they are not easily managed in picture, with their accompaniments of blue sky; but how glorious are they in nature! How pregnant with imagination for the poet! and the height of the Cumbrian mountains is sufficient to exhibit daily and hourly instances of those mysterious attachments. Such clouds, cleaving to their stations, or lifting up suddenly their glittering heads from behind rocky barriers, or hurrying out of sight with speed of the sharpest edge —will often tempt an inhabitant to congratulate himself on belonging to a country of mists and clouds and storms, and make him think of the blank sky of Egypt, and the cerulean vacancy of Italy, as an unanimated and even a sad spectacle. The atmosphere, however, as in every country subject to much rain, is frequently unfavourable to landscape, especially when keen winds succeed the rain, which are apt to produce coldness, spottiness, and an unmeaning or repulsive detail in the distance—a sunless frost, under a canopy of leaden and shapeless clouds, is, as far as it allows things to be seen, equally disagreeable.

It has been said that in human life there are moments worth ages. In a more subdued tone of sympathy may we affirm, that in the climate of England there are, for the lover of nature, days which are worth whole months,—I might say—even years. One of these favoured days sometimes occurs in spring time, when that soft air is breathing over the blossoms and new-born verdure which inspired Buchanan with his beautiful Ode to the First of May; the air, which, in the luxuriance of his fancy, he likens to that of the golden age,—to that which gives motion to the funereal cypresses on the banks of Lethe;—to the air which is to salute beatified spirits when expiatory fires shall have consumed the earth with all her habitations. But it is in autumn that days of such affecting influence most frequently intervene;—the atmosphere seems refined, and the sky rendered more crystalline, as the vivifying heat of the year abates; the lights and shadows are more delicate; the colouring is richer and more finely harmonized; and, in this season of stillness, the ear being unoccupied, or only gently excited, the sense of vision becomes more susceptible of its appropriate enjoyments. A resident in a country like this which we are treating of, will agree with me, that the presence of a lake is indispensable to exhibit in perfection the beauty of one of these days; and he must have experienced, while looking on the unruffled waters, that the imagination, by their aid, is carried into recesses of feeling otherwise impenetrable. The reason of this is, that the heavens are not only brought down into the bosom of the earth, but that the earth is mainly looked at, and thought of, through the medium of a purer element. The happiest time is when the equinoctial gales are departed; but their fury may probably be called to mind by the sight of a few shattered boughs, whose leaves do not differ in colour from the faded foliage of the stately oaks from which these relics of the storm depend: all else speaks of tranquillity;—not a breath of air, no restlessness of insects, and not a moving object perceptible—except the clouds gliding in the depths of the lake, or the traveller passing along, an inverted image, whose motion seems governed by the quiet of a time, to which its archetype, the living person, is, perhaps, insensible:—or, it may happen, that the figure of one of the larger birds, a raven or a heron, is crossing silently among the reflected clouds, while the voice of the real bird, from the element aloft, gently awakens in the spectator the recollection of appetites and instincts, pursuits and occupations, that deform and agitate the world,—yet have no power to prevent nature from putting on an aspect capable of satisfying the most intense cravings for the tranquil, the lovely, and the perfect, to which man, the noblest of her creatures, is subject.

Thus far of climate, as influencing the feelings through its effect on the object of sense. We may add, that whatever has been said upon the advantages derived to these scenes from a changeable atmosphere, would apply, perhaps still more forcibly, to their appearance under the varied solemnities of night. Milton, it will be remembered, has given a *clouded* moon to Paradise itself. In the night season, also, the narrowness of the vales, and comparative smallness of the lakes, are especially adapted to bring surrounding objects home to the eye and to the heart. The stars, taking their stations above the hill tops, are contemplated from a spot like the Abyssinian recess of Rasselas, with much more touching interest than they are likely to excite when looked at from an open country with ordinary undulations: and it must be obvious, that it is the *bays* only of large lakes that can present such contrasts of light and shadow as those of smaller dimensions display from every quarter. A deep contracted valley, with diffused waters, such a valley and plains, level and wide as those of Chaldæa, are the two extremes in which the beauty of the heavens and their connexion with the earth are most sensibly felt. Nor do the advantages I have been speaking of imply here an exclusion of the aerial effects of distance. These are insured by the height of the mountains, and are found, even in the narrowest vales, where they lengthen in perspective, or act (if the expression may be used) as telescopes for the open country.

The subject would bear to be enlarged upon; but I will conclude this section with a night-scene suggested by the vale of Keswick. The fragment is well known, but it gratifies me to insert it, as the writer was one of the first who led the way to a worthy admiration of this country.

> "Now sunk the sun, now twilight sunk, and night
> Rode in her zenith; not a passing breeze
> Sigh'd to the grove, which in the midnight air
> Stood motionless, and in the peaceful floods

Inverted hung, for now the billows slept
Along the shore, nor heav'd the deep; but spread
A shining mirror to the moon's pale orb,
Which, dim and waning, o'er the shadowy cliffs,
The solemn woods, and spiry mountain tops,
Her glimmering faintness threw: now every eye
Oppress'd with toil, was drown'd in deep repose,
Save that the unseen shepherd in his watch,
Propp'd on his crook, stood listening by the fold,
And gaz'd the starry vault and pendant moon;
Nor voice, nor sound, broke on the deep serene;
But the soft murmur of swift-gushing rills,
Forth issuing from the mountain's distant steep,
(Unheard till now, and now scarce heard) proclaim'd
All things at rest, and imag'd the still voice
Of quiet, whispering in the ear of night."*

---

* Dr. Brown, the author of this fragment, was, from his infancy, brought up in Cumberland, and should have remembered that the practice of folding sheep by night is unknown among these mountains, and that the image of the Shepherd upon the watch is out of its place, and belongs only to countries with a warmer climate, that are subject to ravages from beasts of prey. It is pleasing to notice a dawn of imaginative feeling in these verses. Tickle, a man of no common genius, chose, for the subject of a Poem, Kensington Gardens, in preference to Derwent, within a mile or two of which he was born. But this was in the reign of Queen Anne, or George the first. Progress must have been made in the interval, though the traces of it, except in the works of Thompson and Dyer, are not very obvious.

SECTION SECOND.

ASPECT OF THE COUNTRY, AS AFFECTED BY ITS INHABITANTS.

HITHERTO I have chiefly spoken of the features by which Nature has discriminated this country from others. I will now describe, in general terms, in what manner it is indebted to the hand of man. What I have to notice on this subject will emanate most easily and perspicuously from a description of the ancient and present inhabitants, their occupations, their condition of life, the distribution of landed property among them, and the tenure by which it is holden.

The reader will suffer me here to recall to his mind the shape of the valleys, their position with respect to each other, and the forms and substance of the intervening mountains. He will people the valleys with lakes and rivers; the coves and sides of the mountains with pools and torrents; and will bound half of the circle which we have contemplated, by the sands of the sea, or by the sea itself. He will conceive that, from the point upon which he stood, he looks down upon this scene before the country had been penetrated by any inhabitants;—to vary his sensations, and to break in upon their stillness, he will form to himself an image of the tides visiting and revisiting the friths, the main sea dashing against the bolder shore, the rivers pursuing their course to be lost in the mighty mass of waters. He may see or hear in fancy the winds sweeping over the lakes, or piping with a loud voice among the mountain peaks; and, lastly, may think of the primeval woods shedding and renewing their leaves with no human eye to notice, or human heart to regret or welcome the change. "When the first settlers entered this region (says an animated writer) they found it overspread with wood; forest trees—the fir, the oak, the ash, and the birch, had skirted the fells, tufted the hills, and shaded the valleys, through centuries of silent solitude; the birds and beasts of prey reigned over the meeker species; and the *bellum inter omnia* maintained the balance of nature in the empire of beasts."

Such was the state and appearance of this region when the aboriginal colonists of the Celtic tribes were first driven or drawn towards it, and became joint tenants with the wolf, the boar, the wild bull, the red deer, and the leigh, a gigantic species of deer which has long been extinct; while the inaccessible crags were occupied by the falcon, the raven, and the eagle. The inner parts were too secluded, and of too little value, to participate much in the benefit of Roman manners; and though these conquerors encouraged the Britons to the improvement of their lands in the plain country of Furness and Cumberland, they seem to have had little connexion with the mountains, except for military purposes, or in subservience to the profit they drew from the mines.

When the Romans retired from Great Britain, it is well known that these mountain-fastnesses furnished a protection to some unsubdued Britons, long after the more accessible and more fertile districts had been seized by the Saxon or Danish invader. A few, though distinct, traces of Roman forts or camps, as at Ambleside and upon Dunmallet, and a few circles of rude stones attributed to the Druids,* are the only vestiges that remain upon the surface of the country of these ancient occupants; and as the Saxons and Danes, who succeeded to the possession of the villages and hamlets which had been established by the Britons,

* It is not improbable that these circles were once numerous, and that many of them may yet endure in a perfect state, under no very deep covering of soil. A friend of the Author, while making a trench in a level piece of ground not far from the banks of the Eamont, but in no connexion with that river, met with some stones which seemed to him formally arranged: this excited his curiosity, and, proceeding, he uncovered a perfect circle of stones, from two to three or four feet high, with a *sanctum sanctorum*,—the whole a complete place of Druidical worship of small dimensions, having the same sort of relation to Stonehenge, Long Meg and her Daughters neer the river Eden, and Karl Lofts near Shap (if this last be not Danish), that a rural chapel bears to a stately church, or to one of our noble cathedrals. This interesting little monument having passed, with the field in which it was found, into other hands, has been destroyed. It is much to be regretted, that the striking relic of antiquity at Shap has been in a great measure destroyed also.

The Daughters of Long Meg are placed not in an oblong, as the Stones of Shap, but in a perfect circle, eighty yards in diameter, and seventy-two in number, and from above three yards high to less than so many feet: a little way out of the circle stands Long Meg herself—a single stone eighteen feet high.

When the Author first saw this monument, he came upon it by surprise, therefore might over-rate its importance as an object; but he must say, that though it is not to be compared with Stonehenge, he has not seen any other remains of those dark ages which can pretend to rival it in singularity and dignity of appearance.

A weight of awe not easy to be borne
Fell suddenly upon my spirit, cast
From the dread bosom of the unknown past,
When first I saw that sisterhood forlorn;—
And Her, whose strength and stature seem to scorn
The power of years—pre-eminent, and placed
Apart, to overlook the circle vast.
Speak, Giant-mother! tell it to the Morn,
While she dispels the cumbrous shades of night;
Let the Moon hear, emerging from a cloud,
When, how, and wherefore, rose on British ground
That wondrous Monument, whose mystic round
Forth shadows, some have deemed, to mortal sight
The inviolable God that tames the proud.

seem at first to have confined themselves to the open country,—we may descend at once to times long posterior to the conquest by the Normans, when their feudal polity was regularly established. We may easily conceive that these narrow dales and mountain sides, choked up as they must have been with wood, lying out of the way of communication with other parts of the Island, and upon the edge of a hostile kingdom, could have little attraction for the high-born and powerful; especially as the more open parts of the country furnished positions for castles and houses of defence, sufficient to repel any of those sudden attacks which, in the then rude state of military knowledge, could be made upon them. Accordingly, the more retired regions (and to such I am now confining myself) must have been neglected or shunned even by the persons whose baronial or signioral rights extended over them, and left doubtless, partly as a place of refuge for outlaws and robbers, and partly granted out for the more settled habitation of a few vassals following the employment of shepherds or woodlanders. Hence these lakes and inner valleys are unadorned by any remains of ancient grandeur, castle, or monastic edifices, which are only found upon the skirts of the country, as Furness Abbey, Calder Abbey, the Priory of Lanercost, Gleaston Castle, —long ago a residence of the Flemings,—and the numerous ancient castles of the Cliffords, the Lucys, and the Dacres. On the southern side of these mountains (especially in that part known by the name of Furness Fells, which is more remote from the borders), the state of society would necessarily be more settled; though it was also fashioned, not a little, by its neighbourhood to a hostile kingdom. We will, therefore, give a sketch of the economy of the Abbots in the distribution of lands among their tenants, as similar plans were doubtless adopted by other Lords, and as the consequences have affected the face of the country materially to the preset day, being, in fact, one of the principal causes which give it such a striking superiority, in beauty and interest, over all parts of the island.

"When the Abbots of Furness," says an author before cited, "enfranchised their villains, and raised them to the dignity of customary tenants, the lands, which they had cultivated for their lord, were divided into whole tenements; each of which, besides the customary annual rent, was charged with the obligation of having in readiness a man completely armed for the king's service on the borders, or elsewhere; each of these whole tenements was again subdivided into four equal parts; each villain had one; and the party-tenant contributed his share to the support of the man of arms, and of other burdens. These divisions were not properly distinguished; the land remained mixed; each tenant had a share through all the arable and meadow land, and common of pasture over all the wastes. These sub-tenements were judged sufficient for the support of so many families; and no further division was permitted. These divisions and subdivisions were convenient at the time for which they were calculated: the land so parcelled out was, of necessity, more attended to, and the industry was greater, when more persons were to be supported by the produce of it. The frontier of the kingdom, within which Furness was considered, was in a constant state of attack and defence; more hands, therefore, were necessary to guard the coast, to repel an invasion from Scotland, or make reprisals on the hostile neighbour. The dividing the lands in such manner as has been shown, increased the number of inhabitants, and kept them at home till called for: and, the land being mixed, and the several tenants uniting in equipping the plough, the absence of the fourth man was no prejudice to the cultivation of his land, which was committed to the care of three.

"While the villains of Low Furness were thus distributed over the land, and employed in agriculture, those of High Furness were charged with the care of flocks and herds, to protect them from the wolves which lurked in the thickets, and in winter to browze them with the tender sprouts of hollies and ash. This custom was not till lately discontinued in High Furness; and holy-trees were carefully preserved for that purpose when all other wood was cleared off; large tracts of common being so covered with these trees, as to have the appearance of a forest of hollies. At the shepherd's call, the flocks surrounded the holly-bush, and received the croppings at his hand, which they greedily nibbled up, bleating for more. The Abbots of Furness enfranchised these pastoral vassals, and permitted them to enclose *quillets* to their houses, for which they paid encroachment rent."—West's *Antiquities of Furness*.

However desirable, for the purpose of defence, a numerous population might be, it was not possible to make at once the same numerous allotments among the untilled valleys, and upon the sides of the mountains, as had been made in the uncultivated plains. The enfranchised shepherd, or woodlander, having chosen there his place of residence, builds it of sods, or of the mountain-stone, and, with the permission of his lord, encloses, like Robinson Crusoe, a small croft or two immediately at his door, for such animals as he wishes to protect. Others are happy to imitate his example, and avail themselves of the same privileges: and thus a population mainly of Danish or Norse origin, as the dialect indicates, crept on towards the more secluded parts of the valleys. Chapels, daughters of some distant church, are first erected in the more open and fertile vales, as those of Bowness and Grasmere, offsets from Kendal: which again, after a period, as the settled population increases, become mother churches to smaller edifices, planted, at length, in almost every dale throughout the country. The inclosures, formed by the tenantry, are for a long time confined to the homesteads; and the arable and meadow land of the vales is possessed in common field: the several portions being marked out by stones, bushes, or trees; which portions, where the custom has survived, to this day are called *dales*, from the word *deylen*, to distribute; but, while the valley was thus lying open, inclosures seem to have taken place upon the sides of the mountains; because the land there was not intermixed, and was of little comparative value; and, therefore, small opposition would be made to its being appropriated by those to whose habitations it was contiguous. Hence the singular appearance which the sides of many of these mountains exhibit, intersected, as they are, almost to the summit, with stone walls. When first erected, these stone fences must have little disfigured the face of the country, as part of the lines would everywhere be hidden by the quantity of native wood then remaining; and the lines would also be broken (as they still are) by the rocks which interrupt and vary their course. In the meadows, and in those parts of the lower grounds where the soil had not been sufficiently drained, and could not afford a stable foundation, there, when the increasing value of land, and the inconvenience suffered from intermixed plots of ground in common field, had induced each inhabitant to inclose his own, they were compelled to make the fences of alders, willows, and other trees. These, where the native wood has disappeared, have frequently enriched the valleys with a sylvan appearance; while the intricate intermixture of property has given to the fences a graceful irregularity, which, where large properties are prevalent, and large capitals employed in agriculture, is unknown. This sylvan appearance is heightened by the number of ash trees planted in rows along the quick-fences, and along the walls, for the purpose of browzing the cattle at the approach of winter. The branches are lopped off and strewn upon the pastures; and when the cattle have stripped them of their leaves, they are used for repairing the hedges or for fuel.

We have thus seen a numerous body of Dalesmen creeping into possession of their homesteads, their little crofts, their mountain enclosures; and, finally, the whole vale is visibly divided; except, perhaps, here and there some marshy ground, which, till fully drained, would not repay the trouble of inclosing. But these last partitions do not seem to have been general till long after the pacification of the Borders, by the union of the two crowns, when the cause which had first determined the distribution of land into such small parcels had not only ceased, but likewise a general improvement had taken place in the country, with a correspondent rise in the value of its produce. From the time of the union it is certain that this species of feudal population must rapidly have diminished. That it was formerly much more numerous than it is at present, is evident from the multitude of tenements (I do not mean houses, but small divisions of land) which belonged formerly each to several proprietors, and for which separate fines are paid to the manorial lords at this day. These are often in the proportion of four to one of the present occupants. "Sir Launcelot Threlkeld, who lived in the reign of Henry VII., was wont to say, he had three noble houses, one for pleasure, Crosby, in Westmorland, where he had a park full of deer; one for profit and warmth, wherein to reside in winter, namely, Yanwath, nigh Penrith; and the third, Threlkeld (on the edge of the vale of Keswick), well stocked with tenants to go with him to the wars." But, as I have said, from the union of the two crowns, this numerous vassalage (their services not being wanted) would rapidly diminish; various tenements would be united in one possessor; and the aboriginal houses, probably little better than hovels, like the kraals of savages, or the huts of the Highlanders of Scotland, would fall into decay, and the places of many be supplied by substantial and comfortable buildings, a majority of which remain to this day scattered over the valleys, and are often the only dwellings found in them.

From the time of the erection of these houses, till within the last sixty years, the state of society, though no doubt slowly and gradually improving, underwent no material change. Corn was grown in these vales (through which no carriage-road had yet been made) sufficient upon each estate to furnish bread for each family, and no more: notwithstanding the union of several tenements, the possessions of each inhabitant still being small, in the same field was seen an intermixture of different crops; and the plough was interrupted by little rocks, mostly overgrown with wood, or by spongy places, which the tillers of the soil had neither leisure nor capital to convert into firm land. The storms and moisture of the climate induced them to sprinkle their upland property with outhouses of native stone, as places of shelter for their sheep, where, in tempestuous weather, food was distributed to them. Every family spun from its own flock the wool with which it was clothed; a weaver was here and there found among them; and the rest of their wants were supplied by the produce of the yarn, which they carded and spun in their own houses, and carried to market, either under their arms, or more frequently on pack-horses, a small train taking their way weekly down the valley, or over the mountains to the most commodious town. They had, as I have said, their rural chapel, and of course their minister, clothing or in manner of life in no respect differing from themselves, except on the Sabbath-day; this was the sole distinguished individual among them; every thing else, person and possession, exhibited a perfect equality, a community of shepherds and agriculturists,—proprietors, for the most part, of the lands which they occupied and cultivated.

While the process above detailed was going on, the native forest must have been every where receding; but trees were planted for the sustenance of the flocks in winter,—such was then the rude state of agriculture; and, for the same cause, it was necessary that care should be taken of some part of the growth of the native woods. Accordingly, in Queen Elizabeth's time, this was so strongly felt, that a petition was made to the Crown, praying, "the Blomaries in High Furness might be abolished, on account of the quantity of wood which was consumed in them for the use of the mines, to the great detriment of the cattle." But this same cause, about a hundred years after, produced effects directly contrary to those which had been deprecated. The re-establishment, at that period, of furnaces upon a larger scale, made it the interest of the people to convert the steeper and more stony of the inclosures, sprinkled over with remains of the native forest, into close woods, which, when cattle and sheep were excluded, rapidly sowed and thickened themselves. The reader's attention has been directed to the cause by which tufts of wood, pasturage, meadow, and arable land, with its various produce, are intricately intermingled in the same field; and he will now see, in like manner, how enclosures entirely of wood, and those of cultivated ground, are blended all over the country under a law of similar wildness.

A historic detail has thus been given of the manner in which the hand of man has acted upon the surface of the inner regions of this mountainous country, as incorporated with, and subservient to, the powers and processes of nature. We will now take a view of the same agency—acting, within narrower bounds, for the production of the few works of art and accommodations of life which in so simple a state of society, could be necessary. These are merely habitations of man and cover for beasts, roads and bridges, and places of worship.

And to begin with the COTTAGES. They are scattered over the valleys, and under the hills, and on the rocks; and, even to this day, in the more retired dale, without any intrusion of more assuming buildings;

> Cluster'd like stars some few, but single most,
> And lurking dimly in their shy retreats,
> Or glancing on each other cheerful looks,
> Like separated stars with clouds between.

The dwelling houses, and contiguous outhouses, are, in many

instances, of the colour of the native rock, out of which they have been built; but, frequently the dwelling or fire-house, as it is ordinarily called, has been distinguished from the barn or byer by rough-cast and whitewash, which, as the inhabitants are not hasty in renewing it, in a few years acquires, by the influence of weather, a tint at once sober and variegated. As these houses have been, from father to son, inhabited by persons engaged in the same occupation, yet necessarily with changes in their circumstances, they have received without incongruity additions and accommodations adapted to the need of each successive occupant, who, being for the most part proprietor, was at liberty to follow his own fancy: so that these humble dwellings remind the contemplative spectator of a production of nature, and may (using a strong expression) rather be said to have grown than to have been erected;—to have risen, by an instinct of their own, out of the naked rock—so little is there in them of formality, such is their wildness and beauty. Among the numerous recesses and projections in the walls, and in different stages of their roofs, are seen bold and harmonious effects of contrasted sunshine and shadow. It is a favourable circumstance, that the strong winds, which sweep down the valleys, induced the inhabitants, at a time when the materials for building were easily procured, to furnish many of these dwellings with substantial porches; and such as have not this defence, are seldom unprovided with a projection of two large slates over their thresholds. Nor will the singular beauty of the chimneys escape the eye of the attentive traveller. Sometimes a low chimney, almost upon a level with the roof, is overlaid with a slate, supported on four slendar pillars, to prevent the wind from driving the smoke down the chimney. Others are of a quadrangular shape, rising one or two feet above the roof; which low square is often surmounted by a tall cylinder, giving to the cottage chimney the most beautiful shape in which it is ever seen. Nor will it be too fanciful or refined to remark, that there is a pleasing harmony between a tall chimney of this circular form, and the living column of smoke ascending from it through the still air. These dwellings, mostly built, as it has been said, of rough unhewn stone, are roofed with slates, which were rudely taken from the quarry before the present art of cutting them was understood, and are, therefore, rough and uneven in their surface, so that both the coverings and sides of the houses have furnished places of rest for the seeds of lichens, mosses, and flowers. Hence buildings, which in their very form call to mind the processes of nature, do thus, clothed in part with a vegetable garb, appear to be received into the bosom of the living principle of things as it acts and exists among the woods and fields; and, by their colour and their shape, affectingly direct the thoughts to that tranquil course of nature and simplicity, along which the humble-minded inhabitants have, through so many generations, been led. Add the little garden with its shed for bee-hives, its small bed of pot-herbs, and its borders and patches of flowers for Sunday posies, with sometimes a choice few, too much prized to be plucked; an orchard of proportioned size; a cheese-press, often supported by some tree near the door; a cluster of embowering sycamores for summer shade; with a tall fir through which the winds sing when other trees are leafless; the little rill or household spout murmuring in all seasons;—combine these incidents and images together, and you have the representative idea of a mountain-cottage in this country, so beautifully formed in itself and so richly adorned by the hand of nature.

Till within the last sixty years there was no communication between any of these vales by carriage-roads; all bulky articles were transported on pack-horses. Owing, however, to the population not being concentrated in villages, but scattered, the valleys themselves were intersected as now by innumerable lanes and pathways leading from house to house and from field to field. These lanes, where they are fenced by stone walls, are mostly bordered with ashes, hazels, wild roses, and beds of tall fern, at their base; while the walls themselves, if old, are overspread with mosses, small ferns, wild strawberries, the geranium, and lichens: and, if the wall happen to rest against a bank of earth, it is sometimes almost wholly concealed by a rich facing of stone-fern. It is a great advantage to a traveller or resident, that these numerous lanes and paths, if he be a zealous admirer of nature, will lead him into all the recesses of the country, so that the hidden treasures of its landscapes may by an ever-ready guide, be laid open to his eyes.

Likewise to the smallness of the several properties is owing the great number of bridges over the brooks and torrents, and the daring and graceful neglect of danger or accommodation with which so many of them are constructed, the rudeness of the forms of some, and their endless variety. But when I speak of this rudeness, I must at the same time add, that many of these structures are in themselves models of elegance, as if they had been formed upon principles of the most thoughtful architecture. It is to be regretted that these monuments of the skill of our ancestors, and of that happy instinct by which consummate beauty was produced, are disappearing fast; but sufficient specimens remain* to give a high gratification to the man of genuine taste.

* Written some time ago. The injury done since is more than could have been calculated upon. *Singula de nobis anni prædantur euntes.* This is in the course of things, but why should the genius that directed the ancient architecture of these vales have deserted them? For the bridges, churches, mansions, cottages, and their richly-fringed and flat-roofed outhouses, venerable as the

Travellers who may not have been accustomed to pay attention to things so inobtrusive, will excuse me if I point out the proportion between the span and elevation of the arch, the lightness of the parapet, and the graceful manner in which its curve follows faithfully that of the arch.

Upon this subject I have nothing further to notice, except the PLACES OF WORSHIP, which have mostly a little school-house adjoining*. The architecture of these churches and chapels, where they have not been recently rebuilt or modernised, is of a style not less appropriate and admirable than that of the dwelling-houses and other structures. How sacred the spirit by which our forefathers were directed! The *religio loci* is no where violated by these unstinted, yet unpretending, works of human hands. They exhibit generally a well-proportioned oblong, with a suitable porch, in some instances a steeple tower, and in others nothing more than a small belfry, in which one or two bells hang visibly. But these objects, though pleasing in their forms, must necessarily, more than others in rural scenery, derive their interest from the sentiments of piety and reverence for the modest virtues and simple manners of humble life with which they may be contemplated. A man must be very insensible who would not be touched with pleasure at the sight of the chapel of Buttermere, so strikingly expressing, by its diminutive size, how small must be the congregation there assembled, as it were like one family; and proclaiming at the same time to the passenger, in connection with the surrounding mountains, the depth of that seclusion in which the people live, that has rendered necessary the building of a separate place of worship for so few. A patriot, calling to mind the images of the stately fabrics of Canterbury, York, or Westminster, will find a heart-felt satisfaction in presence of this lowly pile, as a monument of the wise institutions of our country, and as evidence of the all-pervading and maternal care of that venerable Establishment, of which it is,

grange of some old abbey, have been substituted structures, in which baldness only seems to have been studied, or plans of the most vulgar utility. But some improvement may be looked for in future; the gentry *recently* have copied the old models, and successful instances might be pointed out, if I could take the liberty.

* In some places scholars were formerly taught in the church, and at others the school-house was a sort of ante-chapel to the place of worship, being under the same roof: an arrangement which was abandoned as irreverent. It con-continues, however, to this day in Borrowdale. In the parish register of that chapel is a notice, that a youth who had quitted the valley, and died in one of the towns on the coast of Cumberland, had requested that his body should be brought and interred at the foot of the pillar by which he had been accustomed to sit while a school-boy. One cannot but regret that parish registers so seldom contain any thing but bare names; in a few of this country, especially in that of Loweswater, I have found interesting notices of unusual occurrences—characters of the deceased, and particulars of their lives. There is no good reason why such memorials should not be frequent: these short and simple annals would in future ages become precious.

perhaps, the humblest daughter. The edifice is scarcely larger than many of the single stones or fragments of rock which are scattered near it.*

We have thus far confined our observations on this division of the subject to that part of these Dales which runs up far into the mountains.

As we descend towards the open country we meet with halls and mansions, many of which have been places of defence against the incursions of the Scottish borderers; and they not unfrequently retain their towers and battlements. To these houses parks are sometimes attached, and to their successive proprietors we chiefly owe whatever ornament is still left to the country of majestic timber. Through the open parts of the vales are scattered, also, houses of a middle rank between the pastoral cottage and the old hall residence of the knight or esquire. Such houses differ much from the rugged cottages before described, and are generally graced with a little court or garden in front, where may yet be seen specimens of those fantastic and quaint figures which our ancestors were fond of shaping out in yew-tree, holly, or box-wood. The passenger will sometimes smile at such elaborate display of petty art, while the house does not deign to look upon the natural beauty or the sublimity which its situation almost unavoidably commands.

Thus has been given a faithful description, the minuteness of which the reader will pardon, of the face of this country as it was, and has been through centuries, till within the last sixty years. Towards the head of these Dales was found a perfect Republic of Shepherds and Agriculturists, amongst whom the plough of each man was confined to the maintenance of his own family, or for the occasional accommodation of his neighbour.† Two or three cows furnished each family with milk and cheese. The chapel was the only edifice that presided over these dwellings, the supreme head of this pure commonwealth; the members of which existed, in the midst of a powerful empire, like an ideal society, or an organized community whose constitution had been imposed and regulated by the mountains which had protected it.

* Since this was written, a new chapel has been erected, on the site of the old one, at the expence of the Rev. T. Vaughan.

† One of the most pleasing characteristics of manners, in secluded and thinly-peopled districts, is a sense of the degree in which human happiness and comfort are dependent on the contingency of neighbourhood. This is implied by a rhyming adage common here, "*Friends are far, when neighbours are nar*" (near). This mutual helpfulness is not confined to out-door work; but is ready upon all occasions. Formerly if a person became sick, especially the mistress of a family, it was usual for those of the neighbours who were more particularly connected with the party by amicable offices, to visit the house, carrying a present! This practice, which is by no means obsolete, is called *owning* the family, and is regarded as a pledge of a disposition to be otherwise serviceable in a time of disability and distress.

Neither high-born nobleman, knight, or esquire, was here; but many of these humble sons of the hills had a consciousness that the land which they walked over and tilled had for more than five hundred years been possessed by men of their name and blood; and venerable was the transition, when a curious traveller, descending from the heart of the mountains, had come to some ancient manorial residence in the more open parts of the vales, which, through rights attached to its proprietor, connected the almost visionary mountain republic he had been contemplating, with the substantial frame of society as existing in the laws and constitution of a mighty empire.

SECTION THIRD.

CHANGES, AND RULES OF TASTE FOR PREVENTING THEIR BAD EFFECTS.

SUCH, as hath been said, was the appearance of things till within the last sixty years. A practice, denominated Ornamental Gardening, was at that time becoming prevalent over England. In union with an admiration of this sort, and in some instances in opposition to it, had been generated a relish for select parts of natural scenery: and Travellers, instead of confining their observations to Towns, Manufactories, or Mines, began (a thing till then unheard of) to wander over the island in search of sequestered spots, distinguished, as they might accidentally have learned, for the sublimity or beauty of the forms of Nature there to be seen. Dr. Brown, the celebrated Author of the "Estimate of the Manners and Principles of the Times," published a letter to a friend, in which the attractions of the Vale of Keswick were delineated with a powerful pencil, and the feeling of a genuine enthusiast. Gray, the Poet, followed: he died soon after his forlorn and melancholy pilgrimage to the Vale of Keswick, and the record left behind him of what he had seen and felt in this journey, excited that pensive interest with which the human mind is ever disposed to listen to the farewell words of a man of genius. The journal of Gray feelingly showed how the gloom of ill health and low spirits had been irradiated by objects, which the Author's powers of mind enabled him to describe with distinctness and unaffected simplicity. Every reader of this journal must have been impressed with the words which conclude his notice of the Vale of Grasmere:—"Not a single red tile, no flaring gentleman's house or garden-wall, breaks in upon the repose of this little unsuspected paradise; but all is peace, rusticity, and happy poverty, in its neatest and most becoming attire."

What is here so justly said of Grasmere applied almost equally to all its sister Vales. It was well for the undisturbed pleasure of the Poet that he had no forebodings of the change which was soon to take place; and it might have been hoped that these words, indicating how much the charm of what *was* depended upon what was *not*, would of themselves have preserved the ancient franchises of this and other kindred mountain retirements from trespass; or (shall I dare to say?) would have secured scenes so consecrated from profanation. The Lakes had now become celebrated; visitors flocked here from all parts of England; the fancies of some were smitten so deeply, that they became settlers; and the Islands of Derwentwater and Windermere, as they offered the strongest temptation, were the first places seized upon, and were instantly defaced by the intrusion.

The venerable wood that had grown for centuries round the small house called St. Herbert's Hermitage, had indeed some years before been felled by its native proprietor, and the whole island planted anew with Scotch firs, left to spindle up by each other's side—a melancholy phalanx, defying the power of the winds, and disregarding the regret of the spectator, who might otherwise have cheated himself into a belief that some of the decayed remains of those oaks, the place of which was in this manner usurped, had been planted by the Hermit's own hand. The sainted spot, however, suffered comparatively little injury. At the bidding of an alien improver, the Hind's Cottage, upon Vicar's Island, in the same lake, with its embowering sycamores and cattle-shed, disappeared from the corner where they stood; and right in the middle, and upon the precise point of the island's highest elevation, rose a tall square habitation, with four sides exposed, like an astronomer's observatory, or a warren-house reared upon an eminence for the detection of depredators, or, like the Temple of Æolus, where all the winds pay him obeisance. Round this novel structure, but at a respectful distance, platoons of fir were stationed, as if to protect their commander when weather and time should somewhat have shattered his strength. Within the narrow limits of this island were typified also the state and strength of a kingdom, and its religion as it had been, and was,—for neither was the Druidical circle uncreated, nor the church of the present establishment; nor the stately pier, emblem of commerce and navigation; nor the fort to deal out thunder upon the approaching invader. The taste of a succeeding proprietor rectified the mistakes as far as was practicable, and has rid the spot of its puerilities. The church, after having been docked of its steeple, is applied, both ostensibly and really, to the purpose for which the body of the pile was actually erected, namely, a boat-house; the fort is demolished; and, without indignation on the part of the spirits of the ancient Druids who officiated at the circle upon the opposite hill, the mimic arrangement of stones, with its *sanctum sanctorum*, has been swept away.

The present instance has been singled out, extravagant as it is, because, unquestionably, this beautiful country has, in numerous other places, suffered from the same spirit, though not clothed exactly in the same form, nor active in an equal degree. It will be sufficient here to utter a regret for the changes that have been made upon the principal Island at Windermere, and in its neighbourhood. What could be more unfortunate than the taste that suggested the paring of the shores, and surrounding with an embankment this spot of ground, the natural shape of which was so beautiful! An artificial appearance has thus been given to the whole, while infinite varieties of minute beauty have been destroyed. Could not the margin of this noble island be given back to nature? Winds and waves work with a careless and graceful hand: and, should they in some places carry away a portion of the soil, the trifling loss would be amply compensated by the additional spirit, dignity, and loveliness, which these agents and the other powers of nature would soon communicate to what was left behind. As to the larch plantations upon the main shore,—they who remember the original appearance of the rocky steeps, scattered over with native hollies and ash trees, will be prepared to agree with what I shall have to say hereafter upon plantations* in general.

But, in truth, no one can now travel through the more frequented tracts without being offended, at almost every turn, by an introduction of discordant objects, disturbing that peaceful harmony of form and colour which had been through a long lapse of ages most happily preserved.

All gross transgressions of this kind originate, doubtless, in a feeling natural and honourable to the human mind, viz. the pleasure which it receives from distinct ideas, and from the perception of order, regularity, and contrivance. Now, unpractised minds receive these impressions only from objects that are divided from each other by strong lines of demarkation; hence the delight with which such minds are smitten by formality and harsh contrast. But I would beg of those who are eager to create the means of such gratification, first carefully to study what already exists; and they will find, in a country so lavishly gifted by nature, an abundant variety of forms marked out with a precision that will satisfy their desires. Moreover, a new habit of pleasure will be formed opposite to this, arising out of the perception of the fine gradations by which in nature one thing passes away into another, and the boundaries that constitute individuality disappear in one instance only to be revived elsewhere under a more alluring form. The hill of Dunmallet, at the foot of Ullswater, was once divided into different portions, by avenues of fir-trees, with a green and almost perpendicular lane descending down the steep hill through each avenue: contrast this quaint

* These are disappearing fast, under the management of the present proprietor, and native wood is resuming its place.

appearance with the image of the same hill overgrown with self-planted wood,—each tree springing up in the situation best suited to its kind, and with that shape which the situation constrained or suffered it to take. What endless melting and playing into each other of forms and colours does the one offer to a mind at once attentive and active; and how insipid and lifeless, compared with it, appear those parts of the former exhibition with which a child, a peasant perhaps, or a citizen unfamiliar with natural imagery, would have been most delighted!

The disfigurement which this country has undergone has not, however, proceeded wholly from the common feelings of human nature, which have been referred to as the primary sources of bad taste in rural imagery; another cause must be added, that has chiefly shown itself in its effects upon building. I mean a warping of the natural mind occasioned by a consciousness, that this country being an object of general admiration, every new house would be looked at and commented upon either for approbation or censure. Hence all the deformity and ungracefulness that ever pursue the steps of constraint or affectation. Persons who, in Leicestershire or Northamptonshire, would probably have built a modest dwelling like those of their sensible neighbours, have been turned out of their course; and, acting a part, no wonder if, having had little experience, they act it ill. The craving for prospect, also, which is immoderate, particularly in new settlers, has rendered it impossible that buildings, whatever might have been their architecture, should in most instances be ornamental to the landscape, rising as they do from the summits of naked hills in staring contrast to the snugness and privacy of the ancient houses.

No man is to be condemned for a desire to decorate his residence and possessions. Feeling a disposition to applaud such an endeavour, I would shew how the end may be best attained. The rule is simple. With respect to grounds: work, where you can, in the spirit of nature, with an invisible hand of art. Planting, and a removal of wood, may thus, and thus only, be carried on with good effect; and the like may be said of building, if Antiquity, who may be styled the co-partner and sister of Nature, be not denied the respect to which she is entitled. I have already spoken of the beautiful forms of the ancient mansions of this country, and of the happy manner in which they harmonize with the forms of nature. Why cannot such be taken as a model, and modern internal convenience be confined within their external grace and dignity. Expense to be avoided, or difficulties to be overcome, may prevent a close adherence to this model; still, however, it might be followed to a certain degree, in the style of architecture and in the choice of situation, if the thirst for prospect were mitigated by those considerations of comfort, shelter, and convenience, which used to be chiefly sought after. But should an aversion to old fashions unfortunately exist, accompanied with a desire to transplant into the cold and stormy North, the elegances of a villa formed upon a model taken from countries with milder climate, I will adduce a passage from an English poet, the divine Spenser, which will shew in what manner such a plan may be realized without injury to the native beauty of these scenes.

> Into that forrest farre they thence him led,
> Where was their dwelling in a pleasant glade
> With MOUNTAINS round about environed,
> And MIGHTY WOODS which did the valley shade,
> And like a stately theatre it made,
> Spreading itself into a spacious plaine;
> And in the midst a little river plaide
> Emongst the puny stones which seem'd to 'plaine
> With gentle murmure that his course they did restraine.
>
> Beside the same a dainty place there lay,
> Planted with myrtle trees and laurels green,
> In which the birds sang many a lovely lay
> Of God's high praise, and of their sweet loves teene,
> As it an earthly paradise had beene;
> In whose *enclosed shadow* there was pight
> A fair pavillion, *scarcely to be seen*,
> The which was all within most richly dight,
> That greatest princess living it mote well delight.

Houses or mansions suited to a mountainous region should be "not obvious, not obtrusive, but retired;" and the reasons for this rule, though they have been little adverted to, are evident. Mountainous countries, more frequently and forcibly than others, remind us of the power of the elements, as manifested in winds, snows, and torrents, and, accordingly, make the notion of exposure very unpleasing; while shelter and comfort are in proportion necessary and acceptable. Far-winding valleys difficult of access, and the feelings of simplicity habitually connected with mountain retirements, prompt us to turn from ostentation, as a thing there eminently unnatural and out of place. A mansion, amidst such scenes, can never have sufficient dignity or interest to become principal in the landscape, and to render the mountains, lakes, or torrents, by which it may be surrounded, a subordinate part of the view. It is, I grant, easy to conceive that an ancient castellated building, hanging over a precipice, or raised upon an island or the peninsula of a lake, like that of Kilchurn Castle, upon Loch Awe, may not want, whether deserted or inhabited, sufficient majesty to preside for a moment in the spectator's thoughts over the high mountains among which it is embosomed; but its titles are from antiquity—a power readily submitted to upon occasion as the vicegerent of Nature: it is respected, as having owed its existence to the necessities of things, as a monument of security in times of disturbance and danger long passed away,—as a record of the pomp and violence of passion, and a symbol of the wisdom of law;—it bears a countenance of authority, which is not impaired by decay.

> "Child of loud-throated war, the mountain-stream
> Roars in thy hearing; but thy hour of rest
> Is come, and thou art silent in thy age!"

To such honours a modern edifice can lay no claim; and the puny efforts of elegance appear contemptible, when, in such situations, they are obtruded in rivalship with the sublimities of Nature. But, towards the verge of a district like this of which we are treating, when the mountains subside into hills of moderate elevation, or in an undulating or flat country, a gentleman's mansion may, with propriety, become a principal feature in the landscape; and, being itself a work of art, works and traces of artificial ornament may without censure, be extended around it, as they will be referred to the common centre, the house; the right of which to impress within certain limits a character of obvious ornament, will not be denied, where no commanding forms of nature dispute it, or set it aside. Now, to a want of the perception of this difference, and to the causes before assigned, may chiefly be attributed the disfigurement which the Country of the Lakes has undergone from persons who may have built, demolished, and planted, with full confidence that every change and addition was, or would become, an improvement.

The principle that ought to determine the position, apparent size, and architecture of a house, viz. that it should be so constructed, and (if large) so much of it hidden as to admit of its being gently incorporated into the scenery of nature—should also determine its colour. Sir Joshua Reynolds used to say, "If you would fix upon the best colour for your house, turn up a stone, or pluck up a handful of grass by the roots, and see what is the colour of the soil where the house is to stand, and let that be your choice." Of course this precept, given in conversation, could not have been meant to be taken literally. For example, in Low Furness, where the soil, from its strong impregnation with iron, is universally of a deep red, if this rule were strictly followed, the house also must be of a glaring red; in other places it must be of a sullen black; which would only be adding annoyance to annoyance. The rule, however, as a general guide, is good; and, in agricultural districts, where large tracts of soil are laid bare by the plough, particularly if (the face of the country being undulating) they are held up to view, this rule, though not to be implicitly adhered to, should never be lost sight of;—the colour of the house ought, if possible, to have a cast or shade of the colour of the soil. The principle is, that the house must harmonize with the surrounding landscape: accordingly, in mountainous countries, with still more confidence may it be said, "look at the rocks and those parts of the mountains where the soil is visible, and they will furnish a safe direction." Nevertheless, it will often happen that the rocks may bear so large a proportion to the rest of the landscape, and may be of such a tone of colour, that the rule may not admit, even here, of being implicitly followed. For instance, the chief defect in the colouring of the Country of the Lakes (which is most strongly felt in the summer season), is an over-prevalence of a bluish tint, which the green of the herbage, the fern, and the woods, does not sufficently counteract. If a house, therefore, should stand where this defect prevails, I have no hesitation in saying, that the colour of the neighbouring rocks would not be the best that could be chosen. A tint ought to be introduced approaching nearer to those which, in the technical language of painters, are called *warm:* this, if happily selected, would not disturb, but would animate, the landscape. How often do we see this exemplified upon a small scale by the native cottages, in cases where the glare of white-wash has been subdued by time and enriched by weather-stains! No harshness is then seen; but one of these cottages, thus coloured, will often form a central point to a landscape by which the whole shall be connected, and an influence of pleasure diffused over all the objects that compose the picture. But where the cold blue tint of the rocks is enriched by the iron tinge, the colour cannot be too closely imitated; and it will be produced of itself by the stones hewn from the adjoining quarry, and by the mortar, which may be tempered with the most gravelly part of the soil. The pure blue gravel from the bed of the river, is, however, more suitable to the mason's purpose, who will probably insist also that the house must be covered with rough-cast, otherwise it cannot be kept dry. If this advice be taken, the builder of taste will set about contriving such means as may enable him to come the nearest to the effect aimed at.

The supposed necessity of rough-cast to keep out rain in houses not built of hewn stone or brick, has tended greatly to injure English landscape, and the neighbourhood of these lakes especially, by furnishing such apt occasion for whitening buildings. That white should be a favourite colour for rural residences is natural for many reasons. The mere aspect of cleanliness and neatness thus given, not only to an individual house, but, where the practice is general to the whole face of the country, produces moral associations so powerful that, in many minds, they take place of all others. But what has already been said upon the subject of cottages, must have convinced men of feeling and imagination, that a human dwelling of the humblest class may be rendered more deeply interesting to the affections, and far more pleasing to the eye, by other influences than a sprightly tone of colour spread over its outside. I do not, however, mean to deny, that a small white building, embowered in trees, may, in some situations, be a delightful and animating object—in no way injurious to the landscape; but this only where it sparkles from the midst of a thick shade, and in rare and solitary instances; especially if the country be itself rich and pleasing, and abound with grand forms. On the sides of bleak and desolate moors, we are indeed thankful for the sight of white cottages and white houses plentifully scattered, where, without these, perhaps every thing would be cheerless: this is said, however, with hesitation, and with a wilful sacrifice of some higher enjoyments. But I have certainly seen such buildings glittering at sunrise, and in wandering lights, with no common pleasure. The continental traveller also will remember that the convents hanging from the rocks of the Rhine, the Rhone, the Danube, or among the Appenines, or the mountains of Spain, are not looked at with less complacency when, as is often the case, they happen to be of a brilliant white. But this is perhaps owing, in no small degree, to the contrast of that lively colour with the gloom of monastic life, and to the general want of rural residences of smiling and attractive appearance, in those countries.

The objections to white, as a colour, in large spots or masses in landscape, especially in a mountainous country, are insurmountable. In nature, pure white is scarcely ever found but in small objects, such as flowers; or in those which are transitory, as the clouds, foam of rivers, and snow. Mr. Gilpin, who notices this, has also recorded the just remark of Mr. Locke, of N———, that white destroys the *gradations* of distance; and, therefore, an object of pure white can scarcely ever be managed with good effect in landscape painting. Five or six white houses, scattered over a valley, by their obtrusiveness, dot the surface, and divide it into triangles, or other mathematical figures, haunting the eye, and disturbing that repose which might otherwise be perfect. I have seen a single white house materially impair the majesty of a mountain; cutting away, by a harsh separation, the whole of its base below the point on which the house stood. Thus was the apparent size of the mountain reduced, not by the interposition of another object in a manner to call forth the imagination, which will give more than the eye loses: but what had been abstracted in this case was left visible; and the mountain appeared to take its beginning, or to rise, from the line of the house, instead of its own natural base. But, if I may express my own individual feeling, it is after sunset, at the coming on of twilight, that white objects are most to be complained of. The solemnity and quietness of nature at that time are always marred, and often destroyed, by them. When the ground is covered with snow, they are of course inoffensive; and in moonshine they are always pleasing—it is a tone of light with which they accord: and the dimness of the scene is enlivened by an object at once conspicuous and cheerful. I will conclude this subject with noticing, that the cold slaty colour, which many persons who have heard the white condemned have adopted in its stead, must be disapproved of for the reason already given. The flaring yellow runs into the opposite extreme, and is still more censurable. Upon the whole, the safest colour, for general use, is something between a cream and a dust colour, commonly called stone colour;—there are, among the Lakes, examples of this that need not be pointed out.*

The principle taken as our guide, viz. that the house should be so formed, and of such apparent size and colour, as to admit of its being gently incorporated with the works of nature, should also be applied to the management of the grounds and plantations, and is here more urgently needed; for it is from abuses in this department, far more even than from the introduction of exotics in architecture (if the phrase may be used), that this country has suffered. Larch and fir plantations have been spread, not merely with a view to profit, but in many instances for the sake of ornament. To those who plant for profit, and are thrusting every other tree out of the way, to make room for their favourite, the larch, I would utter first a regret, that they should have selected these lovely vales for their vegetable manufactory, when there is so much barren and irreclaimable land in the neighbouring moors, and in other parts of the island, which might have been had for this purpose at a far cheaper rate. And I will also beg leave to represent to them, that they ought not to be carried away by flattering promises from the speedy growth of this tree; because in rich soils and sheltered situations, the wood, though it thrives fast, is full of sap, and of little value: and is, likewise, very subject to ravage from the attacks of insects, and from blight. Accordingly, in Scotland, where planting is much better understood, and carried on upon an incomparably larger scale than among us, good soil and sheltered situations are appropriated to the oak, the ash, and other deciduous trees; and the larch is now generally confined to barren and exposed ground. There the plant, which is a hardy one, is of slower growth; much less liable to injury; and the timber is of better quality.

* A proper colouring of houses is now becoming general. It is best that the colouring material should be mixed with the rough-cast, and not laid on as a *wash* afterwards.

But the circumstances of many permit, and their taste leads them, to plant with little regard to profit; and there are others, less wealthy, who have such a lively feeling of the native beauty of these scenes, that they are laudably not unwilling to make some sacrifices to heighten it. Both these classes of persons, I would entreat to enquire of themselves wherein that beauty which they admire consists. They would then see that, after the feeling has been gratified that prompts us to gather round our dwelling a few flowers and shrubs, which, from the circumstance of their not being native, may, by their very looks remind us that they owe their existence to our hands, and their prosperity to our care; they will see that, after this natural desire has been provided for, the course of all beyond has been predetermined by the spirit of the place. Before I proceed, I will remind those who are not satisfied with the restraint thus laid upon them, that they are liable to a charge of inconsistency, when they are so eager to change the face of that country, whose native attractions, by the act of erecting their habitations in it, they have so emphatically acknowledged. And surely there is not a single spot that would not have, if well managed, sufficient dignity to support itself, unaided by the productions of other climates, or by elaborate decorations which might be becoming elsewhere.

Having adverted to the feelings that justify the introduction of a few exotic plants, provided they be confined almost to the doors of the house; we may add, that a transition should be contrived, without abruptness, from these foreigners to the rest of the shrubs, which ought to be of the kinds scattered by Nature through the woods—holly, broom, wild-rose, elder, dogberry, white and black thorn, &c.—either these only, or such as are carefully selected in consequence of their being united in form, and harmonising in colour with them, especially with respect to colour, when the tints are most diversified, as in autumn and spring. The various sorts of fruit-and-blossom-bearing trees usually found in orchards, to which may be added those of the woods,—namely, the wilding, black-cherry tree, and wild cluster-cherry (here called heck-berry), may be happily admitted as an intermediate link between the shrubs and forest trees; which last ought almost entirely to be such as are natives of the country. Of the birch, one of the most beautiful of the native trees, it may be noticed, that in dry and rocky situations, it outstrips even the larch, which many persons are tempted to plant merely on account of the speed of its growth. The Scotch fir is less attractive during its youth than any other plant; but, when full-grown, if it has had room to spread out its arms, it becomes a noble tree; and, by those who are disinterested enough to plant for posterity, it may be placed along with the sycamore near the house; for, from their massiveness, both these trees unite well with buildings, and in some situations with rocks also; having in their forms and apparent substances, the effect of something intermediate betwixt the immoveableness and solidity of stone, and the spray and foliage of the lighter trees. If these general rules be just, what shall we say to whole acres of artificial shrubbery and exotic trees among rocks and dashing torrents, with their own wild wood in sight—where we have the whole contents of the nurseryman's catalogue jumbled together—colour at war with colour, and form with form?—among the most peaceful subjects of Nature's kingdom, everywhere discord, distraction, and bewilderment! But this deformity, bad as it is, is not so obtrusive as the small patches and large tracts of larch plantations that are overrunning the hill sides. To justify our condemnation of these, let us again recur to Nature. The process by which she forms woods and forests is as follows. Seeds are scattered indiscriminately by winds, brought by waters, and dropped by birds. They perish or produce, according as the soil and situation upon which they fall are suited to them: and under the same dependence, the seedling or the sucker, if not cropped by animals (which Nature is often careful to prevent by fencing it about with brambles or other prickly shrubs), thrives, and the tree grows, sometimes single, taking its own shape without constraint, but for the most part compelled to conform itself to some law imposed upon it by its neighbours. From low and sheltered places, vegetation travels upwards to the more exposed; and the young plants are protected, and to a certain degree fashioned, by those that have preceded them. The continuous mass of foliage which would be thus produced, is broken by rocks, or by glades or open places, where the browzing of animals has prevented the growth of wood. As vegetation ascends, the winds begin also to bear their part in moulding the forms of the trees; but, thus mutually protected, trees, though not of the hardiest kind, are enabled to climb high up the mountains. Gradually, however, by the quality of the ground, and by increasing exposure, a stop is put to their ascent; the hardy trees only are left: those also by little and little, give way—and a wild and irregular boundary is established, graceful in its outline, and never contemplated without some feeling, more or less distinct, of the powers of Nature by which it is imposed.

Contrast the liberty that encourages, and the law that limits, this joint work of nature and time, with the disheartening necessities, restrictions, and disadvantages, under which the artificial planter must proceed, even he whom long observation and fine feeling have best qualified for his task. In the first place, his trees, however well chosen and adapted to their several situations, must generally start all at the same time; and this necessity would of itself prevent that fine connexion of parts, that sympathy and organization, if I may so express myself, which pervades the whole of a natural wood, and appears to the eye in its single trees, its masses of foliage, and their various colours, when they are held up to view on the side of a mountain; or when, spread over a valley, they are looked down upon from an eminence. It is therefore impossible, under any circumstances, for the artificial planter to rival the beauty of nature. But a moment's thought will shew that, if ten thousand of this spiky tree, the larch, are stuck in at once upon the side of a hill, they can grow up into nothing but deformity; that while they are suffered to stand, we shall look in vain for any of those appearances which are the chief sources of beauty in a natural wood.

It must be acknowledged that the larch, till it has outgrown the size of a shrub, shews, when looked at singly, some elegance in form and appearance, especially in spring, decorated as it then is, by the pink tassels of its blossoms; but, as a tree, it is less than any other pleasing; its branches (for *boughs* it has none) have no variety in the youth of the tree, and little dignity even when it attains its full growth; *leaves* it cannot be said to have, consequently neither affords shade nor shelter. In spring the larch becomes green long before the native trees; and its green is so peculiar and vivid, that, finding nothing to harmonize with it, wherever it comes forth a disagreeable speck is produced. In summer, when all other trees are in their pride, it is of a dingy lifeless hue; in autumn of a spiritless unvaried yellow; and, in winter it is still more lamentably distinguished from every other deciduous tree of the forest, for they seem only to sleep, but the larch appears absolutely dead. If an attempt be made to mingle thickets, or a certain proportion of other forest trees, with the larch, its horizontal branches intolerantly cut them down as with a scythe, or force them to spindle up to keep pace with it. The terminating spike renders it impossible that the several trees, where planted in numbers, should ever blend together so as to form a mass or masses of wood. Add thousands to tens of thousands, and the appearance is still the same—a collection of separate individual trees, obstinately presenting themselves as such; and which, from whatever point they are looked at, if but seen, may be counted upon the fingers. Sunshine or shadow has little power to adorn the surface of such a wood; and the trees not carrying up their heads, the wind raises among them no majestic undulations. It is indeed true that, in countries where the larch is a native, and where, without interruption it may sweep from valley to valley and from hill to hill, a sublime image may be produced by such a forest in the same manner as by one composed of any other single tree, to the spreading of which no limits can be assigned; for sublimity will never be wanting where the sense of innumerable multitude is lost in, and alternates with, that of intense unity; and to the ready perception of this effect, similarity and almost identity of individual form and monotony of colour contribute. But this feeling is confined to the native immeasurable forest; no artificial plantation can give it.

The foregoing observations will, I hope, (as nothing has been condemned or recommended without a substantial reason,) have some influence upon those who plant for ornamant merely. To such as plant for profit, I have already spoken. Let me then entreat that the native deciduous trees may be left in complete possession of the lower ground; and that plantations of larch, if introduced at all, may be confined to the highest and most barren tracts. Interposition of rocks would there break the dreary uniformity of which we have been complaining; and the winds would take hold of the trees, and imprint upon their shapes a wildness congenial to their situation.

Having determined what kind of trees must be wholly rejected, or at least very sparingly used, by those who are unwilliug to disfigure the country; and having shewn what kinds ought to be chosen; I should have given, if my limits had not already been overstepped, a few practical rules for the manner in which trees ought to be disposed in planting. But to this subject I should attach little importance, if I could succeed in banishing such trees as introduced deformity, and could prevail upon the proprietor to confine himself either to those found in the native woods, or to such as accord with them. This is, indeed, the main point; for much as these scenes have been injured by what has been taken from them—buildings, trees, and woods, either through negligence, necessity, avarice, or caprice—it is not the removals, but the harsh *additions* that have been made, which are the worst grievance—a standing and unavoidable annoyance. Often have I felt this distinction with mingled satisfaction and regret; for, if no positive deformity or discordance be substituted or superinduced, such is the benignity of Nature, that, take away from her beauty, after beauty and ornament after ornament, her appearance cannot be marred—the scars, if any be left, will gradually disappear before a healing spirit; and what remains will still be soothing and pleasing.—

"Many hearts deplored
The fate of those old trees; and oft with pain
The traveller at this day will stop and gaze
On wrongs which nature scarcely seems to heed:
For sheltered places, bosoms, nooks and bays,
And the pure mountains, and the gentle Tweed,
And the green silent pastures yet remain."

There are few ancient woods left in this part of England upon which such indiscriminate ravage as is here "deplored," could now be committed. But, out of the numerous copses, fine woods might in time be raised, probably without sacrifice of profit, by leaving, at the periodical fellings, a due proportion of the healthiest trees to grown up into timber. This plan has, fortunately, in many instances, been adopted; and they who have set the example are entitled to the thanks of all persons of taste. As to the management of planting with reasonable attention to ornament, let the images of nature be your guide, and the whole secret lurks in a few words; thickets or underwoods—single trees—trees clustered or in groups—groves—unbroken woods, but with varied masses of foliage—glades—invisible or winding boundaries—in rocky districts, a seemly proportion of rock left wholly bare, and other parts half hidden—disagreeable objects concealed, and formal lines broken—trees climbing up to the horizon, and, in some places, ascending from its sharp edge, in which they are rooted, with the whole body of the tree appearing to stand in the clear sky—in other parts, woods surmounted by rocks utterly bare and naked, which add to the sense of height, as if vegetation could not thither be carried, and impress a feeling of duration, power of resistance, and security from change!

The author has been induced to speak thus at length, by a wish to preserve the native beauty of this delightful district, because still further changes in its appearance must inevitably follow, from the change of inhabitants and owners which is rapidly taking place. About the same time that strangers began to be attracted to the country, and to feel a desire to settle in it, the difficulty, that would have stood in the way of procuring situations, was lessened by an unfortunate alteration in the circumstances of the native peasantry, proceeding from a cause which then began to operate, and is now felt in every house. The family of each man, whether *estatesman* or farmer, formerly had a twofold support; first, the produce of his lands and flocks; and, secondly, the profit drawn from the employment of the women and children, as manufacturers; spinning their own wool in their own houses (work chiefly done in the winter season), and carrying it to market for sale. Hence, however numerous the children, the income of the family kept pace with its increase. But, by the invention and universal application of machinery, this second resource has been cut off; the gains being so far reduced as not to be sought after but by a few aged persons disabled from other employment. Doubtless, the invention of machinery has not been to these people a pure loss; for the profits arising from home-manufactures operated as a strong temptation to chose that mode of labour in neglect of husbandry. They also participate in the general benefit which the island has derived from the increased value of the produce of land, brought about by the establishment of manufactures, and by the consequent quickening of agricultural industry. But this is far from making them amends: and now that home-manufactures are nearly done away, though the women and children might, at many seasons of the year, employ themselves with advantage in the fields beyond what they are accustomed to do, yet still all possible exertion in this way cannot be rationally expected from persons whose agricultural knowledge is so confined, and, above all, where there must necessarily be so small a capital. The consequence, then, is—that proprietors and farmers being no longer able to maintain themselves upon small farms, several are united in one, and the buildings go to decay, or are destroyed; and that the lands of the *estatesmen* being mortgaged, and the owners being constrained to part with them, they fall into the hands of wealthy purchasers, who, in like manner, unite and consolidate; and, if they wish to become residents, erect new mansions out of the ruins of the ancient cottages, whose little enclosures, with all the wild graces that grew out of them, disappear. The feudal tenure under which the estates are held, has indeed done something towards checking this influx of new settlers; but so strong is the inclination, that these galling restraints are endured; and it is probable that, in a few years, the country on the margin of the Lakes will fall almost entirely into the possession of gentry, either strangers or natives. It is then much to be wished, that a better taste should prevail among these new proprietors; and, as they cannot be expected to leave things to themselves, that skill and knowledge should prevent unnecessary deviations from that path of simplicity and beauty along which, without design and unconsciously, their humble predecessors have moved. In this wish the author will be joined by persons of pure taste throughout the whole island, who, by their visits (often repeated) to the Lakes in the North of England, testify that they deem the district a sort of national property, in which every man has a right and interest who has an eye to perceive and a heart to enjoy.*

---

* See, for a glaring instance of disfigurement of the country, a barn and cow-shed lately erected on a slip of meadow-ground lying between the river and the road on the banks of the Rothay, under Loughrigg Fell, not far below the bridge where the stream is crossed at the entrance from the south of Rydal village. This building, objectionable as it is from its size and form, is yet much more so on account of its intercepting peculiarly beautiful views of the Valley, both up and down the river. This remark is made thus publicly *merely* with a view to prevent like mischief by other proprietors.—1846.

SECTION FOURTH.

ALPINE SCENES COMPARED WITH CUMBRIAN.

As a resident among the Lakes, I frequently hear the scenery of this country compared with that of the Alps; and therefore a few words shall be added to what has been incidentally said upon that subject.

If we could recall, to this region of the Lakes, the native pine-forests, with which, many hundred years ago, a large portion of the heights was covered; then, during spring and autumn, it might frequently, with much propriety, be compared to Switzerland,—the elements of the landscape would be the same,—one country representing the other in miniature. Towns, villages, churches, rural seats, bridges and roads, green meadows and arable grounds, with their various produce, and deciduous woods of diversified foliage which occupy the vales and lower regions of the mountains, would, as in Switzerland, be divided by dark forests from ridges and round-topped heights covered with snow, and from pikes and sharp declivities imperfectly arrayed in the same glittering mantle: and the resemblance would be still more perfect on those days when vapours, resting upon and floating around the summits, leave the elevation of the mountains less dependent upon the eye than on the imagination. But the pine-forests have wholly disappeared; and only during late spring and early autumn is realized here that assemblage of the imagery of different seasons, which is exhibited through the whole summer among the Alps,—winter in the distance,—and warmth, leafy woods, verdure and fertility at hand, and widely diffused.

Striking out, then, from among the permanent materials of the landscape, that stage of vegetation which is occupied by pine-forests, and, above that, the perennial snows, we have mountains, the highest of which little exceed 3,000 feet, while some of the Alps do not fall short of 14,000 or 15,000, and 8,000 or 10,000 is not an uncommon elevation. Our tracts of wood and water are almost as diminutive in comparison; therefore, as far as sublimity is dependent upon absolute bulk and height, and atmospherical influences in connection with these, it is obvious that there can be no rivalship. But a short residence among the British mountains will furnish abundant proof, that, after a certain point of elevation, viz. that which allows of compact and fleecy clouds settling upon or sweeping over the summits, the sense of sublimity depends more upon form and relation of objects to each other than upon their actual magnitude: and, that an elevation of 3,000 feet is sufficient to call forth in a most impressive degree the creative, and magnifying, and softening powers of the atmosphere. Hence, on the score even of sublimity, the superiority of the Alps is by no means so great as might hastily be inferred;—and, as to the *beauty* of the lower regions of the Swiss mountains, it is noticeable—that, as they are all regularly mown, their surface has nothing of that mellow tone and variety of hues by which mountain turf, that is never touched by the scythe, is distinguished. On the smooth and steep slopes of the Swiss hills, these plots of verdure do indeed agreeably unite their colour with that of the deciduous trees, or make a lively contrast with the dark green pine groves that define them, and among which they run in endless variety of shapes—but this is most pleasing *at first sight;* the permanent gratification of the eye requires finer gradations of tone, and a more delicate blending of hues into each other. Besides, it is only in spring and late autumn that cattle animate by their presence the Swiss lawns; and, though the pastures of the higher regions where they feed during the summer are left in their natural state of flowery herbage, those pastures are so remote, that their texture and colour are of no consequence in the composition of any picture in which a lake of the Vales is a feature. Yet in those lofty regions, how vegetation is invigorated by the genial climate of that country! Among the luxuriant flowers there met with, groves, or forests, if I may so call them, of Monk's-hood are frequently seen; the flower of deep rich blue, and as tall as in our gardens; and this at an elevation where, in Cumberland, Icelandic moss would only be found, or the stony summits be utterly bare.

We have, then, for the colouring of Switzerland, *principally* a vivid green herbage, black woods, and dazzling snows, presented in masses with a grandeur to which no one can be insensible; but not often graduated by nature into soothing harmony, and so ill suited to the pencil, that though abundance of good subjects may be there found, they are not such as can be deemed *characteristic* of the country; nor is this unfitness confined to colour: the forms of the mountains, though many of them in some points of view the noblest that can be conceived, are apt to run into spikes and needles, and present a jagged outline, which has a mean effect transferred to canvass. This must have been felt by the ancient masters; for, if I am not mistaken, they have not left a single landscape, the materials of which are taken from the *peculiar* features of the Alps; yet Titian passed his life almost in their neighbourhood; the Poussins and Claude must have been well acquainted with their aspects; and several admirable painters, as Tibaldi and Luino, were born among the Italian Alps. A few experiments have lately been made by Englishmen, but they only prove that courage, skill, and judgment may surmount any obstacles; and it may be safely affirmed, that they who have done best in this bold adventure, will be the least likely to repeat the attempt. But, though our scenes are better suited to painting than those of the Alps, I should be sorry to contemplate either country in reference to that art, further than as its fitness or unfitness for the pencil renders it more or less pleasing to the eye of the spectator, who has learned to observe and feel, chiefly from Nature herself.

Deeming the points in which Alpine imagery is superior to British too obvious to be insisted upon, I will observe that the deciduous woods, though in many places unapproachable by the axe, and triumphing in the pomp and prodigality of Nature, have, in general,* neither the variety nor beauty which would exist in those of the mountains of Britain, if left to themselves. Magnificent walnut-trees grow upon the plains of Switzerland; and fine trees of that species are found scattered over the hill-sides; birches also grow here and there in luxuriant beauty; but neither these, nor oaks, are ever a prevailing tree, nor can even be said to be common; and the oaks, as far as I had an opportunity of observing, are greatly inferior to those of Britain. Among the interior valleys, the proportion of beeches and pines is so great that other trees are scarcely noticeable; and surely such woods are at all seasons much less agreeable than that rich and harmonious distribution of oak, ash, elm, birch, and alder, that formerly clothed the sides of Snowdon and Helvellyn, and of which no mean remains still survive at the head of Ullswater. On the Italian side of the Alps, chesnut and walnut trees grow at a considerable height on the mountains; but, even there, the foliage is not equal in beauty to the "natural product" of this climate. In fact, the sunshine of the south of Europe, so envied when heard of at a distance, is in many respects injurious to rural beauty, particularly as it incites to the cultivation of spots of ground which in colder climates would be left in the hands of nature, favouring at the same time the culture of plants that are more valuable on account of the fruit they produce to gratify the palate, than for affording pleasure to the eye as materials of landscape. Take, for instance, the Promontory of Bellagio, so fortunate in its command of the three branches of the Lake of Como, yet the ridge of the Promontory itself, being for the most part covered with vines interspersed with olive trees, accords but

* The greatest variety of trees is found in the Valais.

ill with the vastness of the green unappropriated mountains, and derogates not a little from the sublimity of those finely-contrasted pictures to which it is a fore-ground. The vine, when cultivated upon a large scale, notwithstanding all that may be said of it in poetry,* makes but a dull formal appearance in landscape; and the olive tree (though one is loath to say so) is not more grateful to the eye than our common willow, which it much resembles; but the hoariness of hue, common to both, has in the aquatic plant an appropriate delicacy, harmonising with the situation in which it most delights. The same may no doubt be said of the olive among the dry rocks of Attica, but I am speaking of it as found in gardens and vineyards in the North of Italy. At Bellagio, what Englishman can resist the temptation of substituting, in his fancy, for these formal treasures of cultivation, the natural variety of one of our parks—its pastured lawns, coverts of hawthorn, of wild-rose, and honeysuckle, and the majesty of forest trees?—such wild graces as the banks of Derwent-water showed in the time of the Ratcliffes; and Gowbarrow Park, Lowther, and Rydal do at this day.

As my object is to reconcile a Briton to the scenery of his own country, though not at the expense of truth, I am not afraid of asserting that in many points of view our LAKES, also, are much more interesting than those of the Alps; first, as is implied above, from being more happily proportioned to the other features of the landscape; and next, both as being infinitely more pellucid, and less subject to agitation from the winds.† Como (which may perhaps be styled the King of Lakes, as Lugano is certainly the Queen) is disturbed by a periodical wind blowing *from* the head in the morning, and *towards* it in the afternoon. The

* Lucretius has charmingly described a scene of this kind:—

Inque dies magis in montem succedere sylvas
Cogebant, infraque locum concedere cultis:
Prata, lacus, rivos, segetes, vinetaque læta
Collibus et campis ut haberent, atque olearum
Cœrula distinguens inter *plaga* currere posset
Per tumulos, et convalleis, camposque profusa:
Ut nunc esse vides vario distincta lepore
Omnia, quæ pomis intersita dulcibus ornant,
Arbustisque tenent felicibus obsita circum.

† It is remarkable that Como (as is probably the case with other Italian Lakes) is more troubled by storms in summer than in winter. Hence the propriety of the following verses:—

Lari! margine ubique confragoso
Nulli cœlicolum negas sacellum
Picto pariete saxeoque tecto;
Hinc miracula multa navitarum
Audis, nec placido refellis ore,
Sed nova usque paras, Nota vel Euro
Æstivas quatientibus cavernas
Vel surgentis ab Adduæ cubili
Cœco grandinis imbre provoluto.—LANDOR.

magnificent Lake of the Four Cantons, especially its noblest division, called the Lake of Uri, is not only much agitated by winds, but in the night time is disturbed from the bottom, as I was told, and indeed as I witnessed, without any apparent commotion in the air; and when at rest, the water is not pure to the eye, but of a heavy green hue—as is that of all the other lakes, apparently according to the degree in which they are fed by melted snows. If the Lake of Geneva furnish an exception, this is probably owing to its vast extent, which allows the water to deposit its impurities. The water of the English Lakes, on the contrary, being of crystalline clearness, the reflections of the surrounding hills are frequently so lively, that it is scarcely possible to distinguish the point where the real object terminates and its unsubstantial duplicate begins. The lower part of the Lake of Geneva, from its narrowness, must be much less subject to agitation than the higher divisions, and, as the water is clearer than that of the other Swiss lakes, it will frequently exhibit this appearance, though it is scarcely possible in an equal degree. During two comprehensive tours among the Alps, I did not observe, except on one of the smaller lakes, between Lugano and Ponte Tresa, a single instance of those beautiful repetitions of surrounding objects on the bosom of the water, which are so frequently seen here: not to speak of the fine dazzling trembling net-work, breezy motions, and streaks and circles of intermingled smooth and rippled water, which makes the surface of our lakes a field of endless variety. But among the Alps, where every thing tends to the grand and the sublime, in surfaces as well as in forms, if the lakes do not court the placid reflections of land objects, those of first-rate magnitude make compensation, in some degree, by exhibiting those ever-changing fields of green, blue, and purple shadows or lights (one scarcely knows which to name them), that call to mind a sea-prospect contemplated from a lofty cliff.

The subject of torrents and water-falls has already been touched upon; but it may be added, that in Switzerland, the perpetual accompaniment of snow upon the higher regions takes much from the effect of foaming white streams; while, from their frequency, they obstruct each other's influence upon the mind of the spectator; and, in all cases, the effect of an individual cataract, excepting the great Fall of the Rhine at Schaffhausen, is diminished by the general fury of the stream of which it is a part.

Recurring to the reflections from still water, I will describe a singular phenomenon of this kind of which I was an eye-witness.

Walking by the side of Ullswater upon a calm September morning, I saw, deep within the bosom of the lake, a magnificent castle, with towers and battlements; nothing could be more distinct than the whole edifice. After gazing with delight upon it for some time, as upon a work of enchantment, I could not but regret that my previous knowledge of the place enabled me to account for the appearance. It was, in fact, the reflection of a pleasure-house called Lyulph's Tower—the towers and battlements magnified and so much changed in shape as not to be immediately recognized. In the meanwhile the pleasure-house itself was altogether hidden from my view by a body of vapour stretching over it and along the hill side on which it stands, but not so as to have intercepted its communication with the lake; and hence this novel and most impressive object, which, if I had been a stranger to the spot, would, from its being inexplicable, have long detained the mind in a state of pleasing astonishment.

Appearances of this kind, acting upon the credulity of early ages, may have given birth to, and favoured the belief in, stories of subaqueous palaces, gardens, and pleasure-grounds—the brilliant ornaments of romance.

With this *inverted* scene I will couple a much more extraordinary phenomenon, which will show how other elegant fancies may have had their origin, less in invention than in the actual processes of nature.

About eleven o'clock in the forenoon of a winter's day, coming suddenly, in company of a friend, into view of the Lake of Grasmere, we were alarmed by the sight of a newly-created island. The transitory thought of the moment was, that it had been produced by an earthquake or some other convulsion of nature. Recovering from the alarm, which was greater than the reader can possibly sympathize with, but which was shared to its full extent by my companion, we proceeded to examine the object before us. The elevation of this new island exceeded considerably that of the old one, its neighbour; it was likewise larger in circumference, comprehending a space of about five acres; its surface rocky, speckled with snow, and sprinkled over with birch trees; it was divided towards the south from the other island by a narrow frith, and in like manner from the northern shore of the lake: on the east and west it was separated from the shore by a much larger space of smooth water.

Marvellous was the illusion! Comparing the new with the old island, the surface of which is soft, green, and unvaried, I do not scruple to say that, as an object of sight, it was much the more distinct. "How little faith," we exclaimed, "is due to one sense, unless its evidence be confirmed by some of its fellows! What stranger could possibly be persuaded that this, which we know to be an unsubstantial mockery, is *really* so; and that there exists only a single island on this beautiful lake?" At length the appearance underwent a gradual transmutation; it lost its prominence and passed into a glimmering and dim *inversion*, and then totally disappeared;—leaving behind it a clear open area of ice of the same dimensions. We now perceived that this bed of ice, which was thinly suffused with water, had produced the illusion, by reflecting and refracting (as persons skilled in optics would no doubt easily explain) a rocky and woody section of the opposite mountain named Silver-how.

Having dwelt so much upon the beauty of pure and still water, and pointed out the advantage which the Lakes of the North of England have in this particular over those of the Alps, it would be injustice not to advert to the sublimity that must often be given to Alpine scenes, by the agitations to which those vast bodies of diffused water are there subject. I have witnessed many tremendous thunder-storms among the Alps, and the most glorious effects of light and shadow: but I never happened to be present when any lake was agitated by those hurricanes which I imagine must often torment them. If the commotions be at all proportionable to the expanse and depth of the waters, and the height of the surrounding mountains, then, if I may judge from what is frequently seen here, the exhibition must be awful and astonishing.—On this day, March 30, 1822, the winds have been acting upon the small Lake of Rydal as if they had received command to carry its waters from their bed into the sky; the white billows in different quarters disappeared under clouds, or rather drifts of spray, that were whirled along, and up into the air by scouring winds, charging each other in squadrons in every direction, upon the Lake. The spray, having been hurried aloft till it lost its consistency and whiteness, was driven along the mountain tops like flying showers that vanish in the distance. Frequently an eddying wind scooped the waters out of the basin, and forced them upwards in the very shape of an Icelandic Geyser, or boiling fountain, to the height of several hundred feet.

This small Mere of Rydal, from its position, is subject in a peculiar degree to these commotions. The present season, however, is unusually stormy;—great numbers of fish, two of them not less than twelve pounds weight, were a few days ago cast on the shores of Derwentwater by the force of the waves.

Lest, in the foregoing comparative estimate, I should be suspected of partiality to my native mountains, I will support my general opinion by the authority of Mr. West, whose Guide to the Lakes has been eminently serviceable to the tourist for nearly fifty years. The Author, a Roman Catholic Clergyman, had passed much time abroad, and was well acquainted with the scenery of the Continent. He thus expresses himself: "They who intend to make the continental tour should begin here; as it will give, in miniature, an idea of what they are to meet with there, in traversing the Alps and Appenines; to which our northern mountains are not inferior in beauty of line, or variety of summit, number of lakes and transparency of water; not in colouring of rock, or softness of turf; but in height and extent only. The mountains here are all accessible to the summit, and furnish prospects no less suprising, and with more variety than the Alps themselves. The tops of the highest Alps are inaccessible, being covered with everlasting snow, which, commencing at regular heights above the cultivated tracts, or wooded and verdant sides, form indeed the highest contrast in nature. For there may be seen all the variety of climate in one view. To this, however, we oppose the sight of the ocean, from the summits of all the higher mountains, as it appears intersected with promontories, decorated with islands, and animated with navigation." —WEST'S *Guide*, p. 5.

GEOLOGY OF THE LAKE DISTRICT,

IN

Letters

ADDRESSED TO

W. WORDSWORTH, Esq.

BY THE

REV. PROFESSOR SEDGWICK, M.A., F.R.S., &c.

WOODWARDIAN PROFESSOR OF GEOLOGY IN THE UNIVERSITY
OF CAMBRIDGE.

GEOLOGY OF THE LAKE DISTRICT.

LETTER I.

My dear Sir,—In writing these letters, I am only endeavouring to perform a promise, made many years since, when I had the happiness of rambling with you through some of the hills and valleys of your native country. One of your greatest works seems to contain a poetic ban against my brethren of the hammer, and some of them may have well deserved your censures: for every science has its minute philosophers, who neither have the will to soar above the material things around them, nor the power of rising to the contemplation of those laws by which Nature binds into union the different portions of her kingdom. But Geology has now a different form and stature from what she had in earlier days: she is the handmaid of labourers who are toiling, as they believe, for the good of their fellow men: she claims kindred with all the offspring of exact knowledge; and she lends no vulgar help to the loftiest investigations of human thought. To reject her altogether, can only be done consistently by one who shuts his eyes to the light of material science; and this, I know, is no part of your philosophy; for no one has put forth nobler views of the universality of nature's kingdom than yourself. You wish not her provinces to be dissevered, but each of them to contribute to the good of the whole state. You believe, however, and I subscribe to the same creed, that material science is only so far truly good, as it tends to elevate the mind of man; giving him a higher conception of his capacities and duties, and a better power in following them to their proper end.

All nature bears the impress of one great Creative Mind, and all parts of knowledge are, therefore, of one kindred and family. In toiling along the narrow path leading to some favourite object of our search, we may perhaps forget the world without us, and so become bigots in our philosophy; labouring only for our own ends, or at best, for that which may seem but for the good of a sect or party. True philosophy has a loftier and better aim. Truth, of whatever kind, she considers as a part of herself, which she has to bring under the government of her will; and her only end is "the glory of God, and the good of man's estate."

But I must leave these high subjects of speculation, and descend to more homely matters: and in commencing my task I meet with a great difficulty. I wish to convey some general notion of the structure of the Lake District; and it would be an easy task, even within the compass of one letter, to enumerate the successive great rock formations, to explain their order, and to give a short description of them. But in this way my narative would inevitably be so dry and repulsive, that no one but a professed geologist would ever think of reading it, and even such a person would do so with very little profit. I wish to address more general readers—any intelligent traveller whose senses are open to the beauties of the country around him, and who is ready to speculate on such matters of interest as it offers to him. I will therefore endeavour to avoid technical language as far as I am able, and I do not profess to teach, in a few pages, the geology of a most complicated country (for that would be an idle attempt); but rather to open the mind to the nature of the subject, and to point out the right way towards a comprehension of some of its general truths.

The region, I wish in this way to notice, is bounded on the *West*, by the sea-coast extending from the mouth of the Eden to the mouth of the Lune—on the *North*, by the low country bordering the Eden, and stretching from the Solway Frith to the calcareous hills near Brough and Kirkby Stephen—on the *East*, by the chain of calcareous mountains which ranges from the neighbourhood of Settle (through Ingleborough, Whernside, Wildboar Fell, &c.) to Stainmoor—and on the *South*, by Morecambe Bay and the lower part of the valley of the Lune. But in the following short sketch, many tracts comprehended within these boundaries, will be hardly noticed.

By whatever line a good observer enters the region enclosed within the above-mentioned limits, he must be struck with the great contrast between the hills and mountains that are arranged on its outskirts, and those which rise up towards its centre. On the outskirts, the mountains have a dull outline, and a continual tendency to a tabular form: but those of the interior have a much more varied figure, and sometimes present outlines which are peaked, jagged, or serrated. This difference arises partly from the nature of the component rocks, and partly from their position: for the more central mountains are chiefly made up of slaty beds, with different degrees of induration, which are highly inclined, and sometimes nearly vertical: while the outer hills are, with limited exceptions, made up of beds which are slightly inclined, and sometimes nearly horizontal.

Good instances of these facts may be seen at Kendal Fell and Whitbarrow Scar. They may be studied in all their details by one who ascends the water-courses between Ingleton and the caves in Chapel-le-dale—and perhaps still better in the valleys between Clapham and Horton. In all these places, the great beds of limestone at the base of the calcareous mountains, are seen to rest upon the inclined edges of the slates; and there are hundreds of other places on the outskirts of the lake-mountains where we may find a similar arrangement of the beds. One whose attention has been caught by such phenomena, and who has learned to draw the right conclusion from them, has taken the first firm step in Geology; he has learned that the tabular calcareous hills, which surround the country of the lakes, are of a newer date than the slate rocks within it.

But our observer must not rest contented with this conclusion. A study of the slate rocks must soon convince him, that their component beds were deposited by the sea, and were once nearly horizontal;—that great disturbing forces afterwards raised them up, and sometimes twisted them into complicated curves, till at length they permanently settled into their present position—and that some of these effects were brought about before the existence of the overlying beds of limestone.

Should this remark lead him to speculate on the interval of time that may have elapsed between the periods of the two formations he has been considering, he may return to some of those places where they are seen one resting on the other; and he will find that the overlying horizontal beds of limestone are sometimes separated from the contorted or inclined beds of slate, by masses of conglomerate or cemented shingles, containing innumerable abraded fragments and rolled pebbles, derived from the harder beds associated with the slates: and from the condition of the pebbles he may prove that, at the time the conglomerates were formed, many of the ancient slates were as hard and solid as they are at the present day. Hence he will further conclude—that the slate rocks (which contain many regular beds of sea shells and corals) were deposited by the sea during a long lapse of ages —that they were elevated and contorted by great internal movements—that they passed nearly into the solid state in which we find them now—that afterwards, on the outskirts of their elevation, they were ground down into great irregular masses and banks of shingle—and that all this succession of events was complete before the existence of any part of the overlying calcareous chain. Such facts will teach him that he has been studying phenomena which not only indicate *succession*, but were elaborated during *vast intervals* of time.

Again, the previous conclusion may be fortified, by an examination of the *organic remains* which are buried in the slate rocks and the overlying limestone. The indications given by the organic forms prove that there had been a complete change in the animal kingdom, between the epochs of the two formations, for they hardly interchange a single species. However incomprehensible this may be, it never could have been brought about, compatibly with any known operations of nature, without a great change of physical conditions, and a long lapse of ages.

What has been stated requires for its comprehension no previous knowledge of Geology: and any man may make the right observations, and draw the right conclusions from them, when he is once awake to the interest of those phenomena which rise up on every side of him, and seem to court his senses.

But there are other questions belonging to the rudiments of Geology, which I may now touch upon. The world is not as it was when it came from its Maker's hands. It has been modified by many great revolutions, brought about by an inner mechanism of which we very imperfectly comprehend the movements; but of which we gain a glimpse by studying their effects; and there are many causes still acting on the surface of our globe with undiminished power, which are changing, and will continue to change it, so long as it shall last.

No one can carefully examine a mountain chain, without being convinced that all its inequalities have been greatly modified; and that there was a time when many of them had no existence: that many yawning chasms where once closed, and many hollows once filled up by continuous bands of the strata, which still tally, even in their minutest subdivisions, on the opposite sides of a gorge or valley. The calcareous mountains and valleys skirting the lake country, offer the most perfect illustrations of this view: and we learn that these mountains, though unaffected by some of the great physical revolutions which elevated the older slates, have been lifted out of the sea, rent asunder, and worn down into their present forms, by other causes of like kind, but acting at a later period.

I may mention a theory which is not without its advocates even now, and which was once a favourite doctrine with a large school of geologists. This theory assumes, that many of the valleys and great depressions presented by the surface of the earth, have been scooped out simply by the erosion (continued during a countless succession of past ages) of the waters flowing through them. I affirm, in reply—that the erosion of rivers and torrents, however indefinitely continued, could not account for the hollows and inequalities of any one of our mountain chains—that in instances, almost without number, we find streams making their way through clefts and gorges of solid rock, and escaping towards the sea on *one* side of a chain, while nature offers them on easy and uninterrupted line of descent on the *other* side—that the configuration of no high country yet examined is in accordance with this theory—and that, as a general fact, the streams and torrents of our hilly regions have flowed, only during a few thousand years, through the channels in which we now behold them.

The lake mountains offer many beautiful illustrations of this conclusion. Let an observer examine the whole course of any river (such, for example, as the Derwent, the Cocker, the Eamont, the Lune, or the Kent) from its mouth to the last threads of its ramification through the higher elevation of the country. He may first mark the transporting powers of a river in the formation of silt and marsh lands; and the way in which the action of vegetable life, producing great layers of bog earth and turf, combines with this transporting power in raising up and changing the surface of the country. From the marsh lands spreading out on the coast (and perhaps resting on beds of shells like those now living in the sea), he may ascend to the mid region of the river's course, and mark the fertilizing influence of the waters, and the beautiful fringe of country that borders them. He may ascend still higher, and see the torrents wearing out deep grooves and ploughing furrows in the sides of the mountains; bearing gravel and rounded stones to the plains below, and exposing them to the action of the elements. Lastly, he may mark the mounds of rubbish at the foot of all the great precipices, and the fragments of solid rock scattered on the sides of the valley by which he is ascending. Impressed by such phenomena, produced during past ages by the erosion of the elements, he may perhaps begin to lean towards that false theory I have before alluded to.

But other facts must, in their turn, be noticed, which have a most important bearing upon the question in debate. As we ascend the ramifications of a river, we frequently meet with pools of comparatively stagnant water; and sometimes a succession of those tarns and lakes which give so much brightness and beauty to the country here described. Now all these expanses of nearly stagnant water (for this is the homely view in which we must now regard them) are the recipients of the mud and gravel brought down from the neighbouring hills. At every point, where a mountain stream enters a tarn or lake, is accumulated a delta of greater or less extent, which is a chronometer to tell us during what time the transporting agents have been carrying on their work. It would be idle to draw any exact conclusion from such rough indicators of past time; but they all conspire in one story, and tell us in plain terms, that mountain torrents, in the channels where they now flow, have been pushing silt and gravel and blocks of stone before them, only during a few thousand years. Had rivers been playing their present part during an indefinite lapse of ages, not a lake or tarn could, I believe, have existed in Westmorland and Cumberland. The same conclusion is forced on the mind by the valleys of North Wales, and of every hilly country I have yet examined.

Should any one ask, how then were these valleys formed? We may reply—by every great disturbing force which has acted on the crust of the earth since the first deposition of the beds which form the mountains. There has been a long succession of physical revolutions; and to the combined effects of them all, the older rocks must have been more or less exposed. But during the last few thousand years, this part of the world has been almost quiescent, and the pencelling of its outline has only been slightly touched by the erosion of the waters and the gnawing of the elements. Again, we are certain that there have been enormous changes in the relative levels of sea and land. Near the top of Ingleborough, about 2000 feet above the coast level, are beds which were once tranquilly deposited at the bottom of the sea; for they are full of well-preserved shells and corals. The highest parts of Snowdon, are marked by impressions of sea shells; and similar organic spoils have been found, in some distant chains, at five times the height of any English mountains. Such changes of level, howsoever brought about, must have produced an incomparably greater transporting power than is shewn in any ordinary action of the elements. Accordingly, in our own country, we find, heaped on the flanks of the mountains, choking up the valleys, and spreading far and wide along the plains, great masses of alluvial drift, entirely unconnected with any erosion of the existing rivers. We believe that these masses were formed by the sea, during periods when it was changing its level; and we sometimes (at the height of considerably more than 1000 feet) see proofs of the truth of our hypothesis, by finding sea shells of modern species, imbedded in the heaps of incoherent rubbish which have been drifted over the surface.

As far as regards the phenomena just noticed, it is a matter of indifference whether we suppose the sea to have come down from the tops of the mountains, or the mountains to have been pushed up from the bottom of the sea. The latter supposition agrees with the known powers of nature, and I know of no other intelligible cause for a change of oceanic level. Mountains are simply the highest points of elevation, marking the places where subterranean forces have pushed upwards with greatest intensity, or met with least resistance. The first movements would throw the horizontal deposits into a dome-shape; if pushed too far, the outer coating of the dome would crack and burst asunder in different directions according to the conditions of the moving and resisting powers. It might be sometimes in lines diverging from a centre, like the higher valleys of Cumberland. These cracks and fissures, whether formed under the sea, or in the open air, would be the first rudiments of future valleys; and it is obvious that at all future times, the abrading power of water would act with most intensity upon the lines of fracture and the projecting ends of the shattered strata. Combining this remark with the fact, that there have been many great oscillations of the land, and a long succession of geological periods marked and dated by the plainest physical records, we need not wonder that the valleys of Cumberland and Westmorland (traversing as they do some of the oldest rocks which have obtained a known place in the chronicles of the earth) should present phenomena not to be explained by any forces, however long continued, which are now seen to act on the surface of the country.

One who is alive to the interest of the subject I have just touched on, may, when following the coast between Carlisle and Lancaster, or ascending by any one of the valleys towards the higher mountains of the district, find excellent examples both of modern river sediment, and more ancient marine drift. The older gravel often contains blocks of enormous size, bearing witness to the greatness of that power which moved them from their parent seat.

But there are transported bowlders unconnected with any other drifted matter, sometimes many tons in weight, and in positions most strange and difficult to be accounted for. To follow this subject into its details would lead me far beyond the limits of this letter. I will, therefore, almost confine my notice to the travelled blocks of Shap granite; and they have too distinct a mineral structure to be mistaken, in whatever company we may meet with them. The manner in which they have been scattered over the surface may be understood from the following facts:—

1. Setting out from Wasdale Crag, near Shap, (where is the parent rock), they have passed over the steep calcareous ridge that stretches from Orton Scar to Knipe Scar; we find them scattered, far and wide, upon the low country bordering the Eden; many of them have been floated to the height of several hundred feet above that river, against the steep sides of the great Cross Fell ridge; and in one or two places, near Dufton, the blocks almost cover the ground, and have been mistaken for the decomposing surface of a great mass of undisturbed granite.

2. They have been carried towards the East, and many were stranded on the barrier of Stainmoor; but thousands of blocks, some of them several tons in weight, were pushed over that ridge and then scattered over the plains of Yorkshire. Some floated

over the Hambleton hills and were lodged in the valleys near Scarborough: many others were driven over the chalk downs to the coast of Holderness.

3. Bowlders from Wasdale Crag, some of very great size, have descended the valley of the Kent to the head of Morecambe Bay. Such a movement we may comprehend, allowing an adequate propelling force of water. But they are not confined to the sides of the water-courses. They have been floated to the tops of hills, and across great chasms and depressions. They are found, in numbers, on the high hills between Kendal and Sedbergh, in positions they could not have reached without crossing valleys, now at least, several hundred feet in depth. I might here notice the bowlders of granite and other hard rock, which have been drifted from the western valleys of Cumberland over the plains of Lancashire and Cheshire, and to the very tops of the hills between Cheshire and Derbyshire—the gigantic masses of crystalline rock (some of them not less than forty or fifty feet in diameter) which have descended from the sides of Mont Blanc, then crossed the great valley of Switzerland, and afterwards been lodged against the sides, or pushed over the tops, of the Jura chain—and the innumerable Scandinavian bowlders which are scattered over the northern plains of Germany, and the *steppes* of central Russia. But my limits admit of no details, and I will rest my conclusions on facts supplied by the north of England.

Here then is a great difficulty. By what power were these "erratic blocks" scattered over the north of England, and lodged in positions that seem so utterly strange and anomalous? We may readily admit any change in the relative level of land and water; and therefore any propelling power of oceanic currents consequent upon such a change, and necessary to account for the superficial drift that sometimes contains, as before stated, recent marine shells at the height of considerably more than a thousand feet. But no propelling force of water seems capable of driving gigantic bowlders across ravines and valleys, from mountain top to mountain top; yet we want an agent capable of doing this, when we endeavour to account for the phenomena above described.

Late observations on the marine shells derived from the upper portion of the Crag of Norfolk and Suffolk, and other very recent marine deposits on the eastern coast of England, make it probable that during a period not long before the great diluvian drift, our climate was much *colder* than it is at the present day. The appearances on the coast of North America have given rise to the same conclusion; and the labours of M.M. Agassiz, Charpentier, and other Swiss naturalists, have, I think, clearly proved that, just before the historic time, the glaciers of the Alps were far more extended than they are now. If this be true, may we not suppose that, at the same period, some of the highest valleys of England and Scotland were filled with glaciers, and that numberless blocks of stone which had rolled down the mountain sides, or been torn off from the neighbouring precipices, were then packed up in thick-ribbed ice?

No one will, I trust, be so bold as to affirm that an uninterrupted glacier could ever have extended from Shap Fells to the coast of Holderness, and borne along the blocks of granite through the whole distance, without any help from the floating power of water. The supposition involves difficulties tenfold greater than are implied in the phenomenon it pretends to account for. The glaciers descending through the valleys of the higher Alps have an enormous transporting power: but there is no such power in a great sheet of ice expanded over a country without mountains, and at a nearly dead level.

The period of refrigeration (if such indeed there were) had at length an end; and we can hardly conceive any general change of climate without some great oscillation in the water level. Let us then suppose the earth to sink, or the ocean to rise up, so that the coast line may reach our higher valleys, and then currents of the sea may float away the ancient glaciers with their imbedded fragments of rock. In this way we can conceive it possible that blocks of Shap granite may have been stranded on the side of Cross Fell, or floated over the top of Stainmoor and the crest of the Hambleton hills; and dropped, by the gradual melting of the icebergs, on the spots where we now find them. Soon afterwards our island may have gained a condition of equilibrium, and the land may have risen, or the sea descended, to its present level; in which there appears to have been very little change during the period of modern authentic history.

The previous hypothesis is not new. It was first started, forty or fifty years since, to explain the transporting power which had brought away millions of bowlders and fragments of rock from the Scandinavian chain, and scattered them over the plains in the north of Germany, and in Poland and a part of Russia. But it seemed to be entangled in the greatest difficulty, for how were we to find *the ice*, which was the most important part of the machinery? Geological phenomena appeared to indicate a gradual lowering of temperature, from the oldest epoch down to the present period: and hence it was inferred, that in the epoch just before the historic time, the earth must have been warmer than in our days. But no analogy can stand against the direct evidence of facts; and if there has been a period of refrigeration, accompanied by a great oscillation in the level of land and water, the glacial theory will then lend itself readily to the transport of the "erratic blocks," and it involves no supposition which is in antagonism with the known workings of nature. For sea and land have changed their relative levels many times; and icebergs, year by year, do bear away great blocks of stone from the arctic regions, and drop them in the sea many hundred miles from the shores they first started from. But whether the glacial theory truly accounts for all the strange movements of the Shap granite above described, is a question on which I wish not to offer any decided opinion.

One thing at least is certain, that, by whatever cause the "erratic blocks" were floated across our valleys and over our mountains, their dispersion took place at a comparatively recent time. For many of them, though lying bare on the surface, and exposed to all the action of our climate, still clink under the hammer, and hardly show more signs of decay than the granite of an Egyptian obelisk. I see no reason for supposing that the movement of the great bowlders necessarily took place before the existence of the human race. On this question there seems no direct or conclusive evidence leading to one side or the other. We know, indeed, that bowlders, like those above described, are often associated with ancient marine drift, containing bones of mammals of *extinct species* (such as Mammoth, Mastodon, Rhinoceros, Hippopotamus, &c. &c.)—and we belive that no human bones have been found in the old gravel of Europe, except in situations which seem to shew that they were introduced at a more recent date. But allowing the negative conclusion, that no human bones were entombed, along with the extinct mammals, in the old gravel of Europe, it does not thence follow, that the human race was in no other part of the world ever coeval with the Mastodon and the Mammoth. Whatever may become of such a question, the direct evidence remains untouched; and the condition of the travelled bowlders of Shap granite proves that they were not floated away from the hills of Westmorland during any ancient and indefinite period of time long before the creation of our species.

If we have the clearest proofs of great oscillations of sea-level, and have a right to make use of them while we seek to explain some of the latest phenomena of Geology, may we not resonably suppose that, within the period of human history, similar oscillations have taken place in those parts of Asia which were the cradle of our race, and may have produced that destruction among the early families of men, which is described in our sacred books, and of which so many traditions have been brought down to us through all the streams of authentic history?*

Whatever may become of this question, and of some others, which the limits of this letter barely permit me to touch upon, this I will affirm, that among the records of creation discovered to us by the monuments of the earth's crust, we find no chapter more difficult than that which links the past with the present, and leads us up to the historic period, and the beginning of the works of man. Among the older records, we find chapter after chapter of which we can read the characters, and make out their meaning; and as we approach the period of man's creation, our book becomes more clear, and nature seems to speak to us in language so like our own, that we easily comprehend it. But just as we begin to enter on the history of physical changes going on before our eyes, and in which we ourselves bear a part, our chronicle seems to fail us— a leaf has been torn out from nature's record, and the succession of events is almost hidden from our eyes. The strange hypotheses even sober and good observers have been driven to invent, in their endeavours to explain phenomena, which, in the language of geology, happened as yesterday, are but proofs of the difficulty and obscurity of that chapter in the natural history of the earth, which, being the nearest to that describing changes of our own days, one might have expected to have been the most plain and legible.

With this remark I conclude my long letter. In my next I hope to notice the successive deposits of the lake mountains, and the way in which they are related to one another.

A. SEDGWICK.

Cambridge, May 23, 1842.

* There is nothing new in this speculation, which must have offered itself, from time to time, to every geologist who wished to connect the past with the present. Bearing upon this subject, some most striking facts are brought to light in a great work on the structure of the Caucasus, by M. Dubois de Montpereux. I knew nothing of this work when the above letter was first printed, and I can now only refer the reader to a short but excellent analysis of it by Mr. Murchison, in his "Anniversary Address to the Geological Society of London in 1843."

I may add, that this letter (in the form in which it is now printed) was in the hands of the publisher before Mr. Hopkins had read a paper on the structure of the Cumbrian mountains. (See the "Anniversary" Address just quoted, and the "Proceedings of the Geological Society of London," for 1842.) With many of his views I agree: with some I differ: but all the opinions of one who combines the habits of patient observation with the resources of exact science demand our consideration and respect. During the present spring (1843), he has made some new experiments on the movements of ice, which have considerably modified his former views; and seem to prove, that glaciers *may* act as a transporting power on planes of very small inclination. Had the unexpected result of these experiments been known when the above letter was first published, I should have modified one or two sentences—especially the one in which it is stated,—"that there is no such transporting power in a great sheet of ice expanded over a country without mountains, and at a nearly dead level." (*Supra* p. 177.)

LETTER II.

MY DEAR SIR,—In my preceding letter I shortly noticed the external features of the Lake District, the structure of its valleys, the erosion of its surface by the daily action of the elements, the accumulations of alluvial silt and gravel within its area, the heaps of diluvial drift, and the great bowlders which have travelled from the higher mountains far and wide over the North of England. My present object is to convey some notion of the structure of the great mountain masses, and to show how the several parts are fitted one to another. This can only be done after great labour. The cliffs where the rocks are laid bare by the sea, the clefts and fissures in the hills and valleys, the deep grooves through which the waters flow,—all must be in turn examined; and out of much seeming confusion, order will at length appear. We must, in imagination, sweep off the drifted matter that clogs the surface of the ground; we must suppose all the covering of moss and heath and wood to be torn away from the sides of the mountains, and the green mantle that lies near their feet to be lifted up; we may then see the muscular integuments, and sinews, and bones of our mother Earth, and so judge of the part played by each of them during those old convulsive movements whereby her limbs were contorted and drawn up into their present posture. But all these preliminary labours must here be taken for granted, and I must content myself with giving, in the best way I can, a bare outline of the results to which observers have in this way come.

The rock formations in the mountain tracts between the basins of the Eden and the Lune (as defined in my former letter), are divided into the following natural groups:—

1. New red sandstone.
2. Magnesian limestone and conglomerate.
3. The carboniferous series, including the carboniferous or mountain limestone.
4. Old red sandstone.
5. Upper slates of Westmorland, Low Furness, and a part of Yorkshire, based on the limestone of Coniston Water Head.
6. A great deposit of green slate and porphyry, forming some of the highest mountains of Furness Fells, Westmorland, and Cumberland.
7. Skiddaw slate, passing in the heart of Skiddaw forest, into a complicated group of crystalline or 'metamorphic' slates.

As all the preceding groups were deposited under the sea, the highest (No. 1.) must be of the newest, and the lowest (No. 7) of the oldest date. From beneath them all rise great masses of granite and other kinds of crystalline unbedded rock (No. 8) pushed by the force of subterranean fires into the positions where we now find them. But the date of their eruption cannot be made out from their inner structure; and we can only define the epochs of their appearance by the effects they have produced on the more regular aqueous deposits through which they have forced their way.

SECTION NO. 1.

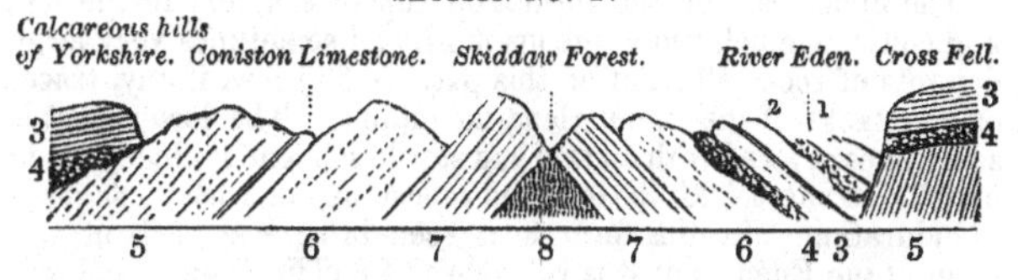

The annexed wood-cut, in which the numerals correspond to those given a little above, may convey some notion of the relative positions of the several great deposits. The left side of the section represents a descending series from the calcareous mountains of Westmorland and Yorkshire to the granite in the centre of Skiddaw forest (No. 8); but some great derangements of the groups, produced by lines of fault, are not delineated, as they would make the section too complicated for a first general view. The right side of the section (commencing with No. 8) represents an ascending series from Skiddaw Forest to Cross Fell. No attempt is however made to give with any exactness the relative magnitudes of the successive groups; nor would it be possible, on such a scale, to delineate the contortions of the beds.

In the order indicated by the numerals, I now proceed to notice the successive formations.

NEW RED SANDSTONE.*

This is the newest formation of the country under notice; for wherever it is associated with other deposits it is always found to rest upon them. It fills all the lower part of the basin of the Eden, from the neighbourhood of Brough to the shores of the Solway Frith. At Maryport it is cut off by the coal measures; but it re-appears at St. Bees' Head, and strikes along the coast to the estuary of the Duddon and the western promontories of Low Furness; and it is seen in a few spots on the shores of Morecambe Bay. In some parts of this long coast range it seems

* No. 1, in the wood-cut.

to have been entirely washed away, and in other places it is covered by enormous heaps of diluvial drift, the colour of which is derived from the abraded fragments of red sandstone.

If we cross to the other side of Morecambe Bay, we meet with the same great formation on the coast of Lancashire; and it may then be traced, through the plains of Cheshire, to the great red central plain stretching across our island from the mouth of the Tees to the mouth of the Severn.

The upper part of the formation supports a very fertile soil, and contains much red gypseous marl, and sometimes very large deposits of rock salt: but of this part we find few, if any, traces on the flanks of the Cumberland mountains. The lower part is sometimes covered with an arid and sterile soil, and is chiefly made up of a strong thickly-bedded red sandstone, in various degrees of induration. In this form it is seen in several parts of the basin of the Eden: but it is valuable as a building stone, and was largely used in the churches and monastic monuments of the middle ages.

The rock here described may be seen, in all its varieties, in the quarries near Carlisle, in the ravines below Furness Abbey, and on the banks of the Calder. At St. Bees' Head it is beautifully exposed to view, and rests on some beds of gypseous marl or 'plaster rock' (not to be confounded with the *upper* gypseous and saliferous marls above noticed), which were formerly much worked. From beneath the gypseous marls rise the magnesian limestone and conglomerate; and these are in their turn underlain by a lower red sandstone, forming a connecting link between the coal series and the deposits I am here enumerating.

The formation seldom appears at a high level. Were England to descend a few hundred feet, all the great central plain above noticed would be under the sea; and the waters of the Solway Frith would extend to the foot of Stainmore, and cover nearly all the space now marked in our geological maps by the colour of the new red sandstone. From this fact we may infer, that the cluster of the Lake mountains and the chain of Cross Fell had been, at least partially, elevated before the period of the new red sandstone. The position of its beds seems to justify this conclusion; for they rest upon the outskirts of the carboniferous rocks in their long range from Kirkby Stephen to Maryport; and, after being expanded on both sides of the Eden, they abut against the great terrace presented by the ridge of Cross Fell (*See woodcut*, p. 181.) A great cleft or 'fault' (sometimes called the 'Pennine fault') ranges from the foot of Stainmore along the base of this terrace, producing such an enormous 'upcast' towards the N. E., that the carboniferous beds, which on one side of the 'fault' are lifted to the height of nearly 3,000 feet, are, on the other side of it, deeply buried underneath the new red sandstone and the alluvion of the Eden. But I must quit a subject requiring for its discussion a knowledge of details I have no right to presume the readers of this letter to be acacquainted with.

Should any one enquire—what was the interval of time between the period of the new red sandstone and of the diluvial rubbish described in the former letter? we may reply, that Cumberland gives us no materials for determining such a question. It only teaches us, that while the drifted matter was forming, the red sandstone was as solid as we now find it in the quarries. This fact, of itself, implies a great interval of time between the two deposits; and other parts of England leave us in no doubt as to the right answer to the previous question.

The new red sandstone is, in many parts of England, overlaid by a series of secondary formations beginning with the lias and ending with the chalk,—each requiring a period of many ages for its elaboration. They contain the remains of many successive creations of organic beings, fitted to perform all the functions of life; but under conditions differing from those of the world in which we now live. Among their strata are the remains of gigantic reptiles,—lines of undisturbed coral reefs,—beds innumerable of sea shells which have lived and died on the spots where we now find them,—and the petrified stumps of trees in the very soil in which they once grew. Phenomena of this kind are repeated again and again. These facts, however striking in themselves, become incomparably more so when studied in combination: and they demonstrate, that successive physical epochs were distinguished by successive changes in the forms of animal and vegetable life,—each change brought about by no natural transmutation of species, nor by any material law we can comprehend, but by an act of Creative Power. However hard it may be for the mind to grasp a succession of facts like these, assuredly long periods of time are implied in their very existence.

Nor do we end here. The chalk and its imbedded flints were all solid, and its organic remains were all petrified before the London clay and the other regular 'tertiary' beds were deposited upon it. The London clay swarms with the traces of organic life, which are utterly unlike the fossils of the chalk, and almost as widely separated from the living *Fauna* of our island. We cannot take one step in Geology without drawing upon the fathomless stores of by-gone time. Man, and all his fellow-beings in the kingdom of animated nature, are creatures but of yesterday: and in no sense (except as the offspring of the same Creating and Controlling Mind) are they the descendants or relations of those beings which are found entombed among the monuments of the ancient world.

But to what does all this tend? It contains a reply to the question before started. Portions of the diluvial drift, and, I believe, all the 'erratic bowlders,' have passed over the country *since* the period of the chalk and of the newest 'tertiary' rocks on the eastern coasts of England. Thousands of ages must therefore have elapsed between the epoch of the new red sandstone and the time of their journey.

There remains another question. If the new red sandstone be of such vast antiquity, what were the forms and conditions of animal and vegetable life coeval with it? The following summary contains the only reply permitted by the limits of this letter.

1. The remains of reptiles appear among the beds of the new red sandstone under forms so strange and anomalous, that anatomists have only found a place for them by interpolating new chapters in nature's history, and separating the class of reptiles into new orders and genera. It contains a lizard with jaws like the beak of a bird of prey; hence the name ***Rhynchosaurus***. In the upper beds of the same formation are impressions of large feet resembling the marks of a human hand; hence the name *Chirotherium*, or hand-beast. These monsters are now proved to be gigantic batrachians (animals of the same order with frogs and toads), and they had jaws armed with formidable teeth resembling those of the crocodile.*

2. The vegetable fossils of the new red sandstone belong to a peculiar ***Flora***. They do not interchange species either with the vegetable fossils of the carboniferous epoch, or with those of the lias and oolites: still less do they resemble the vegetables of the tertiary period, or the present Flora of Europe.

We cannot believe that these successive forms of animated nature were created and destroyed by the mere impulses of a capricious will: but we do believe that they were called into being, and wisely adapted to the successive conditions of our planet, during its progress from a chaotic state till it reached the perfection in which we now find it.

Of the physical changes our planet has undergone, we may gain, at least, a glimmering of knowledge, from a study of its physical records. We may suppose, on analogy, fortified by

* See Professor Owen's admirable "Report on Fossil Reptiles."—*Proceedings of British Association*, 1840 *and* 1841. I forbear to notice in this place the *Ornithichnites* of Professor Hitchcock (impressions of the feet of gigantic birds, &c.) because the exact age of the rocks of red sandstone in which they occur is perhaps not yet determined.

considerations of a more direct and higher kind, that it was once expanded through space in the form of a luminous vapour. We believe, on good evidence, that it was once in a fluid state. The crystalline condition of its inner parts implies a fluidity derived from heat: and if this conclusion be true, the crust of the earth must have passed through many stages of higher temperature before it descended to the mean temperature of the present day. The same conclusion is fortified by the fossils of the older rocks, which indicate a climate warmer than that of the modern period.

Again, enormous masses of carbon are now fixed in the upper parts of the earth's crust, both in chemical combination with other elements, and more simply and tangibly in great beds of coal and other carbonaceous deposits. Much of this fixed and solid carbon may once have floated round the earth as one of the constituents of its atmosphere. A dense atmosphere, highly charged with carbonic acid, may have been well fitted to the rank vegetation of the carboniferous epoch; such an atmosphere may also have been adapted to the respiration of the cold-blooded monsters of the secondary rocks; but utterly unfit for tribes of warm-blooded mammals, created at a later period, and now flourishing on the surface of the earth.

However limited may be our knowledge of the successive physical changes of our planet, this at least is certain, that the Author of Nature has, during all periods, formed organic beings on the same great plan: so that we can reason from the organs to the functions of a cold-blooded monster of the old world, with as much certainty as an anatomist can reason on the adaptation of a skeleton to the habits and wants of a living species.

No sober geologist now dares to give an ideal history of the revolutions of the earth. He may speculate indeed, on points respecting which he is at present supplied with very imperfect evidence: but such speculations he considers of little moment. He studies phenomena, groups them together, contemplates them in all their bearings, and so attempts to rise from phenomena to laws. Should he fail in his first attempts, still all his steps are in the right direction, and in the end will lead him towards some higher truth.*

* In France and Germany the series of rocks above noticed admits of a triple division (called "Trias," or the "Triassic system,") in the following ascending order;—

1. *Gres bigare*, or *Bunter Sandstein*. The equivalent of the new red sandstone of St. Bees' Head and the central plains of England.

2. *Muschelkalk*. A formation altogether wanting in England. Its fossils are very numerous, and form an entirely distinct group: but in their general types they resemble the fossils of the oolitic series more nearly than those of the magnesian limestone or carboniferous rocks.

3. *Marnes irisees*, or *Keuper*. They underlie, and pass into, the Lias, without any apparent break or interruption. The same is true of the gypseous and sa-

MAGNESIAN LIMESTONE AND CONGLOMERATE—LOWER DIVISION OF THE NEW RED SANDSTONE.*

Magnesian limestone and conglomerate.—I have before stated that the magnesian limestone rises from beneath the red marl and sandstone of St. Bees' Head. It is of considerable thickness, and is well exposed in quarries near the roads leading from Whitehaven to St. Bees. To the south of the valley of St. Bees it degenerates into a thin magnesian conglomerate at the base of the sandstone, and afterwards, for many miles further south, the limstone disappears altogether: but it reappears in its characteristic form near the village of Stank, in Low Furness. It is generally of a yellowish brown colour, and of a rather earthy structure, and is often full of cells lined with pure carbonate of lime. In the part of England here described, I believe it contains no organic remains, but many such remains are found in the same rock in its range through Yorkshire and Durham.

At Barrow-mouth, on the north side of St. Bees' Head, the magnesian limestone is seen to rest upon, and to pass into, a conglomerate, or 'pudding stone.' Conglomerates of the same structure, and undoubtedly of the same age, are scattered about the flanks of the hills to the north-east of Whitehaven, (for example, near Gillgarron and Arlecdon, in the lower part of Keskill Beck, near Weddicar Hall, &c.), and are generally lodged in the water-worn hollows and inequalities of the lower red sandstone on which they rest. In some places, however, they rest on the edge of the carboniferous beds without the intervention of any red sandstone.

The conglomerates at Barrow-mouth, under St. Bees' Head, are of insignificant thickness; but at Stenkreth Bridge, near Kirkby Stephen, they are seen in far greater force; and by their unequal resistance to the waters of the Eden have given rise to some very striking scenery. They contain both angular and water-worn fragments of the mountain limestone and coal measures: and we thence infer, that they were not deposited till the carboniferous series had passed into a solid form. It is impossible to study the evidence for this conclusion without being driven to the belief, that a long cycle of ages must have rolled away between the period of the limestone and that of the conglomer-

liferous red marls in many parts of England: from which it follows, that a portion of these marls must represent the *Marnes irises* or *Keuper*. (See Geol: Trans: Lond: Vol. iii. p. 121. Second Series). This conclusion was confirmed by Mr. Strickland and Mr. Murchison, who discovered and described some interesting organic remains from the *Keuper* of Warwickshire. (Geol: Trans: Vol. v. p. 341. Second Series.)

* No. 2, in the wood-cut.

ates which rest upon its edges and are partly made up of its ruins. There are instances without number, in other parts of England, in which the whole new red sandstone series is unconformable to the lower rocks on which it rests. It often passes over their inclined edges, like a lintel over the side-posts of a door; and in such cases we have proof positive, that the lower beds had become solid and were set on edge before the red sandstone was laid upon them.

Lower red sandstone. From beneath the magnesian limestone and conglomerate rises a lower red sandstone, finely exposed in the cliffs on both sides of Whitehaven, and forming a connecting link between the coal series and the deposits above described. In some places it seems to pass by insensible gradations into the true coal measures, and has the mineral structure of a common grey carboniferous sandstone. More frequently, it is of a red tint, or is streaked and variegated with red; and there are many quarries where it cannot be distinguished from the red sandstone of St. Bees' Head. Again, though it may in some places pass into the coal measures, as a more general rule, it is placed in a discordant position on their inclined edges. Such appears to be its more common position in the neighbourhood of Whitehaven; and the same rule holds in the range of the deposits through Yorkshire and Durham.*

We may therefore conclude that, in the North of England, the lower red sandstone is, by its structure and position, more nearly related to the formations above it, than to those below it.

The fossils of the magnesian limestone and the lower red sandstone, point to an opposite conclusion, and will perhaps hereafter induce geologists to separate the formations entirely from the new red sandstone (of St. Bees' Head, &c.) and to consider them as the newest members of the 'palæozoic' or 'transition' class; of which the carboniferous rocks form an integral part. The following facts are all that the limits of this letter permit me to bring forward.

1. In the neighbourhood of Bristol, the magnesian conglomerates contain, though rarely, the remains of reptiles—among them the *Palæosaurus* (old lizard) is of a new genus, approaching, in the structure of its teeth, to the hard-backed crocodiles, but in its general bony structure coming more nearly to the scaly lizards.

* In Warwickshire and Shropshire, the lowest red sandstone is generally conformable to the coal measures: but in a part of the former county, the upper red sandstone and saliferous marls are unconformable to the lower. Traces of this discordancy of position may be found in Cumberland; for the conglomerates at the base of the new red sandstone (St. Bees' Head, &c.) sometimes rest, as above stated, on an uneven water-worn surface of the lower red sandstone.

2. The corals, shells, and fish of the magnesian limestone, with a few exceptions, differ in species from the fossils of the carboniferous series. At the same time there are many generic forms in this limestone identical with those of the older rocks, but unlike any which appear between the lias and the chalk, or in any newer deposits; so that the general zoological type of the magnesian limestone very nearly approaches that of the carboniferous.

3. Vegetable fossils are abundant in the lower red sandstone, and cannot, as a group, be distinguished from those of the carboniferous epoch.

It is obvious that these facts, as far as they go, support the conclusion to which I have pointed.*

CARBONIFEROUS SERIES.†

The rocks included under this name, form an irregular girdle almost surrounding the higher lake mountains. To describe them in detail would require a large volume: and I must content myself with little more than a bare enumeration of the four groups into which they may be conveniently divided.

First group, or *Upper Coal-measures*.—This group extends along the coast from the north side of St. Bees' Head to Maryport; and at both places, as before stated, it is covered by the new red sandstone. From the coast it may be followed to the interior, where it bends round the north side of the higher mountains, gradually diminishing in breadth, and at length ending abruptly in the neighbourhood of Rosley Hill. It contains many thin worthless bands of coal; but there are eight or ten

* It has been contended that the existence of fossil reptiles in the magnesian limestone separates it from the carboniforous epoch, and brings it more nearly into the class of the "Trias" and the oolites. But the force of this objection has been taken away by Mr. Lyell, who has recently shown that reptiles have, in North America, left their traces among rocks containing numerous mountain limestone fossils. Mr. Murchison also states that, in the "Permian system" (a great deposit in Russia, of the age of the magnesian limestone) there is a Flora distinct from that of the carboniferous period. Facts like these are of great importance in questions of classification; but I can hardly attend to them without touching on subjects beyond the aim of these letters.

In England, between the marine deposits of the magnesian and carboniferous limestone, there is an interruption caused by the interpolation of the upper coal series, which is not marine. But during the long epoch of the upper coal measures, marine deposits, to which we have nothing analogous in England, *must* have been formed in other parts of the world. In Russia, all the deposits are marine, from the base of the carboniferous, to the higher beds of the "Permian" system. Does it not, therefore, follow that there *may* be marine deposits in Russia intermediate between the magnesian and carboniferous limestones of this country? The answer to this question may perhaps tend to reconcile the conflicting opinions lately given by Mr. Lyell and Mr. Murchison respecting the exact age of a large class of rocks in North America.

† No. 3, in the wood-cut.

different beds in it which have been profitably worked. The two beds from which coal has been so long extracted near Whitehaven and Workington (one of them, the *main band*, sometimes nine or ten feet thick), are in the upper part of the group; the four or five beds formerly worked in the Harrington field, are in the lower part of it; and its aggregate thickness is perhaps not less than 1,000 feet.

The whole deposit once consisted of alternations of sand and finely laminated mud; with countless fragments of drifted vegetable—sometimes single, sometimes matted together in thick and widely-extended beds. Occasionally the plants are upright in posture, and so entire that they seem not to have been drifted from the spots on which they grew: in such cases the coal-beds become the indications of forests and bogs submerged in by-gone ages during the changes of level between land and water. In course of time the drifted sand-beds became sandstone;—the mud became slaty clay or shale;—the vegetable fossils were bituminized; and the whole formation passed into the condition in which we now see it.

In the upper part of this group (as exhibited in different parts of the North of England) there are no marine remains; but it contains some beds of shells belonging to fresh-water genera. All the plants are of extinct species: many of them of extinct genera; and they are of forms which indicate a high tropical temperature. Among them are coniferous trees, like those in some of the South-sea Islands; gigantic reeds; tree-ferns; enormous creeping plants with sharp pinnated leaves (*Stigmariæ*); trees with fluted stems; and many other strange but beautiful forms of vegetable life, seemingly pushed to rankness and luxuriance by great heat and moisture.

It is in vain to speculate on the exact duration of the carboniferous epoch: but we are sure that it lasted through a vast period of time.

One who has any feeling for the wonders of the old world, and any interest in the powers of human skill, will do well to visit the Whitehaven coal-field. The enormous under-ground excavations—the costly machinery—a living world many hundred feet beneath the surface of the earth—works on a gigantic scale extending far under the bottom of the sea—the streams of gas perpetually rising from the coal-beds, which thus give back to the atmosphere a part of the very elements they once drank up from it—the great breaks and contortions of the solid strata —the prodigious influence the mineral treasures are now exerting upon the habits of the whole civilized world—these assuredly, in whatever light we regard them, physically or morally, are topics of no vulgar interest. But, inviting as the subject is, I must here leave it.

Second Group, or *Millstone Grit*.—This group is of complicated structure, being made up of coarse sandstone (occasionally used for millstones), silicious flagstone, shale, and two or three thin bands of coal. On the north side of the lake mountains it is seen only in a very degenerate form: but in the calcareous chain on the south-eastern side, it is finely exposed to view along the tops of the highest mountains; and is not less than six or seven hundred feet thick. There is not, however, any single mountain in which this whole series is well exhibited. Of the coarser grits, deserving the name of millstone, there are three great beds; the lowest of which forms the tabular rock, resting, like a huge coping stone, on the top of Ingleborough. The coal beds in this group are generally very poor, and only worked by horizontal drifts from the sides of the mountains; but a little above Hawes they increase in thickness, and are worked to considerable profit, by vertical shafts.

Few shells have been found in this subdivision of the carboniferous series; but as it rests upon marine deposits, and in some parts of Yorkshire is surmounted by beds with marine shells, we may conclude that it is of marine rather than of fresh-water origin; in which case we must consider the coal beds as formed by vegetable matter drifted from the land into a shallow sea or estuary.

Third Group, or *Shale Limestone*.—This group forms the upper part of the calcareous zone on the north side of the Cumbrian mountains. There, however, it never rises to a high level, and it is so much covered up with drifted matter, that its subdivisions cannot be easily followed. But in the brows of the higher hills between Penyghent and Stainmoor it is seen in great perfection, and sometimes reaches the thickness of 1,000 or 1,200 feet. To give one example; all the great precipices under the crown of Ingleborough, are made up of the rocks of this complicated group, in which are five beds of limestone, alternating with shale, sandstone, and a few thin bands of coal. The beautiful fossil marble, so much used in the north of England, is derived from the two highest calcareous beds of this group; the black marble is obtained exclusively from the lowest. Several of the caol bands, especially one under the highest (or upper scar) limestone, have been extensively worked, both by horizontal drifts and by shafts. All the limestone beds are full of marine shells and corals: from which we may conclude, that the coal bands, alternating with them, were formed of vegetable matter which had drifted into the sea.

Fourth and lowest Group, or *Great Scar Limestone*.—This beautiful rock is almost entirely made up of animal remains, especially shells and corals; and must once have stretched far and wide among shores and shoals which, though long obliterated from the face of the earth, were the first rudiments of the British

Isles. During this period the scar limestone formed a fringing coral reef round the cluster of the lake mountains. Even now, it may be traced uninterruptedly through the greater part of their circumference; and on the west coast between Egremont and Duddon-mouth, where it has almost disappeared, there are three small patches of limestone seeming to indicate its former continuity on that side of Cumberland.

On the southern limits of the country here described, this great reef was in ancient times severed by faults and breaks, which were gradually opened out into wide valleys: but it requires little effort of imagination to conceive that all the great patches of limestone, now marked in this part of our geological maps, were once united. On the eastern limit of the country under notice, the limestone forms an almost pure and uninterrupted calcareous mass, five or six hundred feet in thickness. In the northern part of the zone it degenerates in thickness, and is interrupted by alternating beds of sandstone.

It must, during the progress of its formation, have been comparatively solid: and hence, during subsequent periods of its disruption and elevation, it was incomparably less contorted than the older slate rocks, which at one time were soft and pliable. To its internal structure, and to all the disturbing forces that have since acted upon it, we are to ascribe its extraordinary features—its mural precipices, its caverns, its reciprocating springs, and its deep clefts and gorges. No formation in our island shows features of more play and beauty. The fair bright islands of Killarney—the clefts of Cheddar, and St. Vincent's rocks—the delicious valleys of the Wye and the High Peak—(and to come nearer the lake country) the sublime gorge of Gordale—the fine grey precipices at the foot of Ingleborough—the caverns of Chapel-le-dale and Clapham—the rocks of Kirkby Lonsdale bridge—and the great white terrace of Whitbarrow—all belong to the features of this limestone.

The organic remains of this rock are in infinite abundance, and are described at great length by many authors, especially by Mr. Phillips.* In this place it is only necessary to state that, considered as a group, they differ specifically from the fossils both of the older and newer formations. The newer deposits, commencing with the new red sandstone, contain, as above mentioned, numberless reptiles, many of which were of gigantic size, and were the tyrants and scavengers of the ancient deep. In the carboniferous series no reptiles have yet been found: their place is supplied by animals of a different class, but of kindred habits—fierce 'sauroid fish'—creatures breathing by the help of gills, and having the skeletons of fish; but with jaws armed with great conical teeth like those of large crocodiles or lizards.

* *Geology of Yorkshire*, Vol. 2.

Though the limestone is, like a great potsherd, broken into many fragments, and is now elevated to the tops of mountains, yet its beds, excepting on the lines of certain great faults, are nearly horizontal in its whole southern and eastern range. In its northern range it is considerably more tilted. The horizontal limestone (as before noticed) is seen to rest on the inclined slate rocks in the valleys between Horton and Clapham, without the intervention of a conglomerate. But in such cases, the jagged edges of the slates have been worn off by the continued erosion of water, and rubbed down almost to a smooth horizontal surface: a fact which shows that there must have been a long interval of time between the elevation of the slates and the commencement of the superincumbent coral reef. At Thornton Force, near Ingleton (a place on every account deserving a visit), the inclined slates are separated from the horizontal limestone by a thin band of conglomerate; and thus we arrive at the same conclusion by independent evidence.

In terminating this notice of the carboniferous series, I may remark, that very thin bands of impure coal are occasionally found in the great scar limestone—that all its darker beds derive their colour from bituminous matter—and that, in a few places within the district, carbonaceous shales appear near its base, and have given rise to unprofitable coal works. But the same dark shales in the range of the series from Stainmoor through Cross Fell towards Scotland, become greatly expanded, and alternate with sandstone; and, at length, in the basin of the Tweed, give rise to a profitable coal field far below the geological level of any one which is worked in the more southern parts of our island.

OLD RED SANDSTONE, ETC.*

This deposit is made up of marl, sandstone, and coarse conglomerate; marking a period of great attrition produced by the beating of the sea upon the edges of the old contorted slates, from their first elevation to the time when the reefs of limestone began to form about them. The older rocks were solid, and had been scooped into deep valleys before the existence of the greater part of the conglomerates. This conclusion is proved by the condition of the imbedded pebbles; and by the fact, that in the upper part of the valley of the Rother, above Sedbergh, enormous masses of the old red conglomerates almost fill up an ancient valley of the slate rocks. It is implied also, though on less impressive evidence, from the position of the conglomerates in the upper parts of the basin of the Kent. The formation is interrupted and irregular; having to all appearance been ground down, by the action of the sea upon the older strata, into great banks

* No 4, in the wood-cut.

of coarse shingles, but never spread out into long and continuous beds. This, at least, is its present appearance between the terrace of the limestone and the slates.

Near Orton, a deposit of a red, and sometimes a grey sandstone, resting upon a conglomerate, seems to form an under terrace to the limestone, but its relations are not clear. In its farther range to the N. W. the formation is almost always seen as a conglomerate; and in that state is shown in three or four places between Shap Fells and the river Lowther; but always under the limestone terrace or near its base. Its largest development is in the very coarse conglomerates near the foot of Ullswater, where it rises into a succession of round-topped hills several hundred feet high, and is of great thickness: but towards the N. W. it suddenly dies away. In the neighbourhood of Hesket Newmarket, it however breaks out again in three or four insignificant patches; after which it is not again seen under the long range of the carboniferous rocks towards the west coast of Cumberland.

There is perhaps no true passage between the old red sandstone above described, and the overlying beds of limestone: it is not, however, probable that any long lapse of time intervened between one formation and the other. As soon as the rude mechanical action that produced the conglomerates had ceased, the shell beds and coral reefs began to skirt the ancient shores.

In Herefordshire and some of the neighbouring counties, the old red sandstone exhibits a complete and uninterrupted sequence of deposits from the slate rocks to the carboniferous limestone, and is of enormous thickness. It has long been divided into three groups—the lowest characterized by red flagstone (or 'tilestone') —the middle group by bands of concretionary limestone (or 'cornstone')—the highest by red sandstone and conglomerate. As a general rule, the old red sandstone of the North of England represents only the highest of these three groups. While the two lower groups were forming in Herefordshire, the active powers of nature were employed, among the Cumbrian mountains, in elevating and contorting the ancient rocks, and not in laying down new deposits. To this remark there seems to be an almost solitary exception on the banks of the Lune, a little above Kirkby Lonsdale: for there we meet with some beds of red flagstone, of the age of the lower beds of 'tilestone' and full of fossils, surmounted by bands of concretionary limestone, and by red marls and conglomerates. But even there the sequence is not complete and uninterrupted: for the red flagstones were in a solid state, and were tilted up, before the marls and conglomerates were formed upon them.*

* Mr. Murchison states that the fish-beds of the old red sandstone of Russia are spread over a surface larger than our island!

We have no right to expect many organic remains in a coarse mechanical rock like that above described. But in Scotland and Herefordshire the formation contains beds with many fossils, especially fish: and of all strange monsters, they are amongst the strangest which underground labours have brought to the light of day. As a group, they differ generically from all other living and fossil fish: some of them, in external characters, making a link with the crustacean order—having the gills and skeleton of a fish combined with a rough bony covering like that of a crab.* In other places, especially in Devonshire, the formation has the mineral structure of a slate rock, and abounds with shells and corals; which, considered as a group, are formed on a type intermediate between that of the carboniferous limestone and of the older slates.†

Before I attempt any shetch of the older slate rocks of the Cumbrian mountains, let me endeavour to translate into common language that chapter in the strange old chronicles of the earth, of which we have been turning over the leaves from the end to the beginning.

First, then, we have the record of an ancient revolution given by the old conglomerates.—*Secondly*, the great scar limestone tells us of a long period of repose. Its coral reefs were formed in a shallow sea (for in such seas only do corals grow): but in course of time it sank down, and a sea many hundred feet deep floated over it, and spread out upon it banks of sand, and mud, and drifted vegetables washed from the neighbouring land.—*Thirdly*, again was a period of repose, when a second bank of limestone, with its shells and corals, was tranquilly deposited; after which was a second subsidence, like the former, and followed by like effects. These operations were six times repeated in the formation of the eastern calcareous mountains; each period of repose and each subsidence producing a repetition of like phenomena.—*Fourthly*, came the period of the millstone grit,

* I take this opportunity of strongly recommending to the reader, a work on the Old Red Sandstone of Scotland, at once popular and scientific, and full of the most lively interest; by Mr. H. Miller, Edinburgh, 1841.

† After an examination of the fossils in the hills between Kendal and the Lune, I found it impossible to separate the "tilestone" from the rocks on which it rests. (*See Proceedings of Geol. Soc. Lond. Nov.* 1841.) Mr. Murchison (*Silurian System*) adopted a subdivision of the Old Red Sandstone which had been some years published, and was suggested by the physical structure of Herefordshire. It does not, however, represent the natural grouping of the fossils; and he would now place the lower part of the "tilestones" in the upper division of his "Ludlow Rock." In the text, I am not however discussing the classification of the rocks of Herefordshire, but endeavouring to give an answer to the question—whether there be any section, among the lake mountains, showing a complete sequence of deposits from the upper slate rocks to the mountain limestone. I have replied in the negative as to the only spot in which the answer admits of any doubt. The "tilestones" of Helm, near Kendal, throw no light upon the question, as they are not overlaid by any newer rock.

when the bays and estuaries were gradually filled up, and marine animals ceased to leave their traces among the waters.—*Lastly*, the lagoons and estuaries were converted into lakes and marshes: a rank tropical vegetation covered the ground, and produced the materials of future coal-fields.

Still we are compelled to invoke the same powers of nature: for some of our coal-fields are thousands of feet in thickness, and I can see no intelligible means of accounting for them without the intervention of vast and repeated changes between the levels of land and water. But here I will escape from the slippery ground of hypothesis, and conclude this long letter.

A. SEDGWICK.

Cambridge, *May* 24, 1842.

LETTER III.

MY DEAR SIR,—In my former letter I described the New red sandstone, the Carboniferous series, and the Old red sandstone skirting the Lake mountains. I must now attempt a sketch of the slate rocks and granitic masses of the central regions. Technical details I wish as far as possible to avoid: but I cannot omit them altogether, and am reluctantly compelled to begin this letter with them.

Among the deposits above described, there is seldom any difficulty in making out the order of the beds: but the slate rocks are highly inclined; sometimes set on edge, occasionally (though rarely in the lake country) turned upside down; so that their order is in certain places involved in almost inextricable confusion. Every one, who pretends to observe for himself, must be provided with a good map and a pocket compass; and as he rambles across the country, he may often see the slaty beds rising like a knife's edge through the soil, and running over the hills and across the valleys in undulating or zig-zag lines. At such points of view, he may, by help of his compass, easily determine, in a general way, the directions of the beds, and the points towards which they incline. Should he wish to make more accurate observations, he must be provided with a *spirit-level*, for determining a horizontal plane, and a *clinometer*, for measuring the inclination of the beds: but these instruments (though easily packed along with the compass in a small pocket-case) are only necessary to one who is engaged in a detailed survey.

The *true direction* of a stratum at any point, is represented by the line formed by the intersection of the smooth surface of the stratum with a horizontal plane, and is determined correctly by the horizontal edge of the spirit-level when applied to the surface. This line is technically called *the strike* of the bed; and a line drawn on the surface of a bed, perpendicular to this line of *strike*, is called the *line of dip* or *rise*, accordingly as we take it in the descending or ascending direction. The quantity of *dip* is measured by the *clinometer*, and gives the inclination of the line of dip to the horizon. The directions of the several lines are determined by the compass. In this way, after multitudes of observations and comparisons (carefully registered, and if possible, laid down on a map), we may make out all the essential changes of *dip* and *strike;* and we gradually learn to connect them together, to explain the features of the country by their help, and to draw from them results that are consistent with one another, and tell us the true order of the mineral masses.

But among the older and more crystalline slates it is sometimes impossible to distinguish the several strata so as to mark their position. All the slate beds were at first in the condition of a very fine mud or silt, deposited, layer upon layer, by the sea: and in passing into a solid state the layers cohered so firmly as to become inseparable afterwards by any ordinary means. But another change of structure was at the same time brought about: the particles all nnderwent a new crystalline arrangement (like that of the laminæ of a piece of spar) producing a regular *cleavage* more or less inclined to the original beds. It is by these *cleavage* planes, and not along the planes of the *true beds*, that the quarry-men obtain the fine roofing slates. The observer must therefore learn to distinguish the nearly vertical laminations of the great open slate quarries from the true beds which are generally much less inclined.*

How then are we to determine the position of the true beds of slate?—This can sometimes be done by help of alternating bands of coarser materials wherein the original bedding has not been obliterated by the slaty structure: a mass of slate between two such bands, must have its bedding parallel to them, whatever may be the direction of its laminæ of cleavage.—In other instances we infer the position of the true beds merely from analogy, knowing their situation in the neighbouring country.—Fortunately we may in many cases ascertain the lines of the true beds by an internal and secure test. The planes of the slates are often marked by parallel stripes of different colours. Among the finer green slates these stripes are generally paler than the other part of the rock; and as they mark the original lines of sediment, they are therefore parallel to the true bedding;

* In Wales, Devonshire, and Cornwall there are many quarries, where the cleavage planes are less inclined than the bed.

indeed they generally mark the passage from one bed to another. Sometimes these stripes are seen on slaty laminæ cutting through pyritous bands with shells and corals; and in such cases the stripes upon the smooth surfaces of the slates are always parallel to the fossil bands.

To make this structure understood, let us place flat layers of coloured clay one over another, and then press them together so that they may cohere and form one plastic mass; and let us so arrange them that no layer of coloured clay may be visible excepting the one at the top. In this position no inner structure can meet the eye; but if a cut be made with a knife vertically through the mass, parallel stripes of colour (representing the different layers of clay) will immediately shew themselves on the face of the section thus obtained. The artificial section made by the knife represents the vertical slaty planes obtained by the quarry-man's wedge; and the stripes of coloured clay are strictly analogous to the sedimentary lines upon the smooth surface of the slates.

There are, however, quarries of coarse slate, or flagstone, without the crystalline structure and the fine even surfaces above described, in which the bedding is distinctly visible, and each flagstone represents a true bed. The ripple mark (exactly like that on sea-sand between high and low water) is sometimes seen on the surface of such beds, and they are occasionly studded with the impressions of organic remains. Many of them are found on the hills south of Kendal, especially on Kirkby Moor; but the finest examples are seen in the quarries near Ingleton and Horton.*

There is another difficulty in the structure of slate rocks which must be shortly noticed. They are often intersected by a double set of parallel fissures or 'joints,' produced apparently by a contraction of the mass while passing into a solid state. These lines may have been influenced by the crystalline action of the whole mass; for they often divide the rocks on a mountain side into regular prismatic blocks, and produce much confusion in the position of the true beds. They do not, however, so affect the inner composition of the rock as to produce persistent laminæ parallel to their own planes; and they are not therefore to be confounded with slaty cleavage. Their direction and inclination is variable; but when they nearly coincide with the strike of the beds they may be called *strike joints;* and when they are nearly transverse to the *strike* they may be called *dip joints.*—I must, however, here quit these dry details. My only wish, in alluding to them, is to save the observer from early difficulties, and to

* The Horton flags have, however, an obscure cleavage plane, which sometimes injures the quality of the stone.

start him in the right direction. After all, it is only by experience in the field that he will learn to interpret correctly the complicated characters impressed upon the older slates.*

UPPER DIVISION OF THE SLATE ROCKS.†

This division is based on the calcareous slates, which stretch from Millum, in the south-western corner of Cumberland, through the head of Coniston Water and the head of Windermere, to the neighbourhood of Shap Wells. To the south of this line, it is expanded through Furness Fells and a considerable portion of Westmorland; being bounded to the south-east by Morecambe Bay and the carboniferous formations above described. The rocks within this area may be separated into several ill-defined groups. Three will be here adopted, in the hope that, as the country is more examined and better understood, they may be brought into strict accordance with the three principal Silurian groups of Mr. Murchison.‡.

Upper Group.—This group commences with red flagstones, which, above Kirkby Lonsdale, and close to their junction with the old red sandstone, contain calcareous concretions and numerous fossils. In making a traverse towards Kirkby Moor, the red flagstone is succeeded, in descending order, by purple, grey,

* Among the Cumbrian mountains, the laminæ of slaty cleavage are generally inclined at a great angle to the horizon. Sometimes the beds undulate and the cleavage planes remain constant. In such cases, the inclination of the cleavage planes to the true beds is continually changing. In Devonshire and Cornwall we find (though very rarely) highly inclined beds with nearly horizontal cleavage planes; and we also find cleavage planes of great perfection which are parallel to the true beds. I know of no examples of like kind in the North of England: for there the cleavage planes (at least in the fine slate quarries) are always transverse to the beds; but amongst the finer slates the *strike* of the beds and the *strike* of the cleavage planes are nearly in the same direction. Again, in Devonshire, Cornwall, and North Wales, and in the chain of the Ardennes, I have seen a second set of cleavage planes, beautifully penetrating the slate rocks, and shewing the perfection of their crystalline arrangement: and these double cleavage planes were associated with the striped and double-jointed structure above noticed. As far as I know, there is no example of a *second cleavage plane* to be seen among the lake mountains; and it is a rare appearance in the countries above noticed.

† No. 5, in the wood-cut.

‡ In a paper read before the Geological Society of London, in 1832, I adopted Mr. J. Otley's threefold division of the Cumbrian slate rocks: and I separated the upper division into three ill-defined groups; viz:—

First, the fossiliferous rocks of the fells south of Kendal, and of Kirkby Moor.—*Secondly*, rocks like the former in structure, but with a more slaty impress, and with very few traces of fossils.—*Thirdly*, a complicated group of calcareous slate (of which there are two principal bands), alternating with hard coarse siliceous beds, and with several thick beds of fine roofing slate obtained by transverse cleavage (Ireleth slate)—the whole resting on the fossiliferous limestone of Coniston Water Head. By "the three principal Silurian groups," are meant all the rocks described under the names "*Ludlow*," "*Wenlock*," *and* "*Caradoc*." The "*Llandeilo flags*" have no distinct representative in the north of England.

greenish grey, and blue flagstone.—Some of the greenish bands exactly resemble the harder flagstones among the 'Ludlow rocks,' of Mr. Murchison: and the red flags nearly resemble the 'tile stones' of Herefordshire, but are far less crystalline and micaceous.

Still in descending order, the flagstones are followed by the hard grey siliceous rocks which extend, with many undulations and changes of 'strike' through the hills between the upper part of the valleys of the Kent and the Lune. Among them are beds with an imperfect slaty structure; and here and there are open and earthy bands (giving an honeycombed appearance to the rock) not unusually of a redish brown colour, and with innumerable casts of fossils. Very thin, impure, calcareous beds (but of no continuity, and unfit for use) are seen in a few places near the lines of fossils. The most remarkable of them is at Oxenholme, on the side of the old road from Kendal to Kirkby Lonsdale.

The whole group appears to be based on a set of hard thick beds, among which the fossils gradually disappear. They are of various colours: blueish-grey, greenish-grey, and occasionally of a dark purple and reddish tint; but their characters and distribution are ill-defined. We may perhaps class with these the hard thick beds which break out from under Kendal Fell, and the similar beds which skirt the marshes near Witherslack and extend to the hills near Lindal.

Middle Group.—This group contains many hard, thick, siliceous beds, nearly like those at the base of the preceding subdivision; but subordinate to it are striped flagstones, coarse slates with a decided transverse cleavage producing the striped surfaces above described. Good examples of this kind may be seen on the road from Kendal to Bowness, and on the old road from Kendal to Newby Bridge.

The fine elevations of Howgill Fells and Middleton Fell are chiefly formed by the rocks of this subdivision: but those mountains are separated from the formations on the west bank of the Lune by enormous 'faults,' and are thrown into such contortions that it is difficult to reduce the subordinate masses to any certain order. Their *strike* also differs from that on the west bank of the Lune, being nearly east and west; and at the north end of Middleton Fell, the beds are so much bent to the south as to range nearly at right angles to the average *strike* of the central mountains. The more slaty beds of this group generally effervesce with acids: but in no part of it have any good fossil bands been yet found. Hence there is considerable uncertainty as to its exact geological place; especially as its upper and lower limits are so ill defined

Lower, or *Ireleth slate, Group.*—The base of this group is well defined by the range of Coniston limestone.* *(See the woodcut.)* Its upper limit is not defined by any fossil bands, and may be considered in some measure as arbitrary: but it must inclose all the calcareous beds, and all the beds of good roofing slate. If a line be drawn from the crest of the hills between Broughton and Ulverston, through the foot of Coniston Water, to a point a little below the Ferry House on Windermere; and from thence be prolonged (bending a little towards the east, so as to preserve a parallelism to the range of the Coniston limestone) through the lower part of Long Sleddale and the contorted slates near the foot of Bannisdale, it may be assumed as an approximate boundary between the lower and middle groups.

Among the deposits on the north side of this line a slaty structure decidedly predominates; and the rocks weather into fine picturesque forms, of which there are many beautiful examples between Broughton and the foot of Coniston Water. The same features on a less scale are seen near the Ferry House on Windermere, where the rocks have an aspect so unlike the higher groups, that I at first mistook their nature, and supposed them to represent some ancient slates brought out by a great dislocation.

The most remarkable beds in this group split, by a transverse cleavage, into fine roofing slates—distinguished from the more ancient slates, chiefly by a darker colour, and by the absence of green chloretic flakes upon the surface of the laminæ. Noble quarries have long been opened in these slates near Kirkby Ireleth. Very fine beds of a dark-coloured flagstone (sometimes superficially coated with crystals of pyrites) are also worked in this group, especially in its lowest portions. It contains also three or four bands of calcareous slates, two of which are fossiliferous. One of these ranges on the south side of the estuary of the Duddon—the other, already noticed, forms the base of the whole series. The latter is the most important from its numerous fossils, its thickness and continuity, and from its enormous shifts and displacements in its long range: especially where it strikes across the valleys that intersect its course. In this way it becomes an indication, not merely of the prevailing *strike*

* As this limestone forms the base of the whole upper division of the slate rocks, it may perhaps be well to give its range in more detail. It is seen at Beck, Water Blain, and Greystone House, in Cumberland. It then crosses into High Furness and may be followed by Broughton Mills, and Applethwaite, &c., to Yew Tree, near Coniston. Thence, after two enormous dislocations, it may be followed over the hills north-east of Coniston to Pool Wyke, near the head of Windermere. From the hills above Low Wood, it may be again followed across Troutbeck, over the hills to Kentmere Hall, and thence to Long Sleddale, where it is exposed in quarries near Little London. Lastly, after being lost under the turf bogs, and partly cut off by the granite, it re-appears near Shap Wells, and so passes under the carboniferous rocks.

of the group, but of the manner in which its mineral masses have been fractured and dissevered during the periods of their elevation.

Before I quit the *upper division* of the slate rocks, I may remark that the prevailing *strike* of the lower group is N. E.; and traces of the same general impress may be found in the two upper groups as far as the shores of Morecambe Bay. There are however some remarkable deviations from this rule even in the lower group; and in the two upper groups the exceptions are so numerous, and the rocks exhibit such complicated undulations, that it is difficult to bring their bearings to any rules of symmetry. Again, the great scar limestone skirting the shores of Morecambe Bay is literally shattered into fragments by enormous north and south 'faults:' and all the slate rocks on the southern border of the lake mountains have also been ripped up by great 'faults' (with the same general direction), which have greatly altered the positions and bearings of the beds. Valleys have been scooped out on lines of fracture: and all the great water channels that descend towards Morecambe Bay (from the Lune on the east to the Duddon on the west) have a prevailing north and south course.

There still remains a question—what is the age of this upper division of the slate rocks? An answer can only be given by an appeal to the fossils. So far as I am acquainted with the fossils of the upper group, they contain about forty species found in the 'Upper Silurian' rocks of Mr. Murchison, and five or six which he formerly referred to the lowest beds of the old red sandstone: but which in Westmorland are distributed through the whole of the upper group. With these, are eight or ten species not yet described. The conclusion is inevitable, viz. that the whole group represents only the Upper Silurian rocks (Ludlow, &c). The Coniston limestone and the calcareous slates of Kirkby Ireleth contain numerous corals of the Wenlock and Dudley limestones. Among them the chain coral (*Catenipora*) is abundant. They contain also one or two Silurian Trilobites; and shells of several *genera* (especially the genus *Orthis*) specifically the same with the shells of the 'Caradoc sandstone.' It therefore follows that the base of the lower group, here described, is of the age of the lower Silurian rocks (not using that term in any extended and indefinite sense, but strictly as it was first employed by its author)—and that the whole upper division represents the 'Silurian system;' the middle part of it being, unfortunately, almost without fossils to help us in the demarcation of the three groups.*

* I profess to make no material changes in the text of these letters, and the above paragraph is reprinted as it was first written. But I have now a much better set of fossils, some of which Mr. Sowerby has been engaged in figuring: and a short account of them will be given in an appendix. I formerly attempted

MIDDLE DIVISION OF THE SLATE ROCKS—GREEN SLATE AND PORPHYRY.*

This division forms a vast group, rising into the highest and most rugged mountains of the whole region. It contains two distinct classes of rock—aqueous and igneous: but they are piled one upon another in tabular masses of such regularity, and are so interlaced and blended, that we are compelled to regard them as the effects of two distinct causes, acting simultaneously during a long geological period. The igneous portions present almost every variety of felstone and felstone porphry; sometimes passing into greenstone, and rarely into masses with a structure like that of basalt. All the aqueous rocks have more or less a slaty structure, and pass in their most perfect form into the finest roofing slates.†

to class the Coniston limestone with the limestone of Bala, and the rocks of the middle group with the slate rocks of the Berwyns and of South Wales. But after the discovery of a better arrangement of the Devonian slates, I abandoned this view, and adopted the one here given. Of the older rocks of South Wales, I know little from personal survey, and there are but few parts of that extensive country which I have ever visited. I believe, however, that its older rocks will be found nearly of the same age with those of Carnarvonshire and Merionethshire. The fossils of the Coniston and Bala limestones are indeed very nearly the same. But I do not wish to bring them into close comparison: because the fossiliferous rocks of North Wales (with a lower Silurian type) are of an enormous thickness; and contain bands of organic remains, some of which are far below, and some, if I mistake not, far above the limestone of Bala. In the Geological Map of Westmorland, belonging to the 'Kendal Natural History Society,' one tint only is given to the upper division of the slates, from the impossibility of drawing, with any degree of correctness, the lines of demarcation between the groups. Should fossils be ever found in sufficient abundance to determine the point, it might be well perhaps to tint the whole division in two colours—one representing the upper and the other the lower Silurian rocks. In the absence of well-defined calcareous beds (Wenlock limestone) any further subdivision will perhaps be found impossible.—Of the Coniston fossils, I procured during my survey a good series, which has been since improved by some excellent specimens I owe to my friend Mr. J. Marshall. My list from the upper group has been greatly improved by the kind assistance of my friends Messrs. Gough and Danby, of Kendal, and by specimens procured from Mr. John Ruthven. My best fossils from Kirkby moor were procured in 1822, under the guidance of Smith, the 'father of English geology,' on the day I first became acquainted with him.

* No. 6, in the wood-cut.

† I have adopted the word *felstone* from the Germans; who, by the word *feldstein*, sometimes express those minerals which we commonly, but inaccurately, have called *compact felspar*. The words *compact spar* involve a contradiction. The name *schaalstein* (or, shale-stone) has been applied to a great variety of slaty rocks, in Nassau and the Hartz, intermediate between true slates and erupted trappean rocks—The word *plutonic* is used to distinguish igneous rocks, erupted under the sea, from *volcanic* rocks which have been poured out in the open air. Any rock is called a *porphry*, which has a nearly uniform base studded with crystals.—*Granite* is formed by the union of *quartz*, *felspar*, and *mica*—when the *mica* is replaced by *hornblende*, the rock becomes a *syenite*.—*Greenstone* is a fine-grained rock composed of *felspar* and *hornblende*, and when these minerals are well defined, the rock is called a *syenitic greenstone*.—When the crystals are very small, and the rock almost compact, it is said to

But why are rocks, so different both in appearance and origin, to be confounded in one formation ?—Because nature has made them inseparable. The tabular masses of true erupted 'plutonic rock' alternate with, and pass by insensible gradations into, great beds of breccia and 'plutonic' silt. The breccias are often as hard as the parent rocks; being cemented by a felspathic paste, occasionally studded with garnets and crystals of felspar; and they sometimes put on a columnar form; and the plutonic silt passes into a hardy, flaky, shining rock, which often has a transverse cleavage with an uneven, shining, wavy surface (exactly like that of some varieties of German *schaalstein*). We have only to follow such changes a little farther, and we are conducted, without seeing where we pass their boundaries, into great deposits of the most perfect roofing slates. Of these slates, quartz in the finest state of comminution, and earthy chlorite partly derived from the plutonic silt, are the chief constituents.

The plutonic rocks were poured out under a deep sea; and the breccias were formed mechanically (like volcanic breccias found among streams of modern lava), and were cemented under great pressure. The plutonic silts have an intermediate structure; but their beds must have been spread out by the waters of the sea. The roofing slates are but the extreme case of fine aqueous sediment, chiefly derived from the erupted matter, and sinking into successive beds during intervals of repose: and so far they are analogous to the fine beds of volcanic silt as often formed by the waters of a lake out of the ashes of a modern crater.

In the Cumbrian mountains, no organic remains are found among these rocks. The aqueous deposits seem to have been too often interrupted by igneous action to permit the growth of shell beds and coral banks. Shells and corals are however found (though rarely) among the slate rocks of Snowdonia: but there the igneous beds are less abundant, and were probably poured out at longer intervals of time.

When I began, twenty years since, to examine the lake country, I believed in the igneous origin of basaltic and porphyritic rocks: but I was staggered in my creed, and filled with astonishment, almost at every step, when I saw the alternating masses of slate and porphyry, and the way in which they were blended together. The Wernerian hypothesis has now passed away, and has been extinguished by the more mature discoveries of an advancing science; but it lent itself readily to the expla-

be *basaltic*.—These different forms of rock pass insensibly one into another.—A *conglomerate* is formed by pebbles more or less rounded by water.—A *breccia* is chiefly made up of angular fragments. All the minerals mentioned in these letters may be easily procured, and will soon be sufficiently familiar to any one who wishes to study the older rocks.

nation of many perplexing facts, and had the merit, at first sight, of great simplicity; and I may venture to affirm, that no one is prepared to understand it, or to do any justice to its author, who has not studied, in the field, such phenomena as are continually offered by the Cumbrian slates.*

The southern boundary of this great group is defined by the range of the Coniston limestone. The northern boundary cannot be well understood without the aid of a geologieal map: but an approximation may be made to it by drawing a line from the foot of Wolf Crag to Wanthwaite Crag—continuing it thence by Wallow Crag, near Keswick—by the foot of the great precipices at the head of Newlands, the base of Honister Crag, and the upper precipices of High Stile, and so round by the great coves of Ennerdale Head to the north side of the Hay Cock—and lastly, from the Hay Cock to the north side of Seatallan, and thence in a devious line, which turns to the north, extending several miles beyond Ponsonby Fells. With limited exceptions, all the stratified rocks (aqueous and igneous) in the high mountains inclosed within these boundaries, strike towards the N.E., and dip at a great angle towards the S.E.; and their whole thickness, after every deduction, must be enormous. The beds were set on edge by a gigantic force, urging them from below; and in the progress of elevation, mountain masses were torn asunder and starred by diverging lines of 'fault.' In a few places, indeed, the dip was reversed; but the great beds of porphry (which must have passed into a solid state in cooling) held the masses firm, and kept them from being twisted and bent about, like the upper slates.

Of the brecciated rocks, above described, a fine example occurs on the side of the road at Barrow near Keswick. Masses, similar in structure and colour, pass through Wanthwaite Crag and the foot of Binsey Crag. Numerous examples may also be seen in the great precipices that overhang the higher parts of Eskdale, Wastdale, Ennerdale, and Borrowdale; in the passes between Borrowdale and Grasmere; at the head of Kentmere; and on almost every line of traverse through the higher mountains.†

The plutonic silt, and other beds intermediate between the erupted rocks and the slates, are spread, here and there, almost through the whole country under notice. They are sometimes cellular (probably from the action of heat), the cells being

* The alternations of aqueous and igneous rocks have been illustrated, with many excellent details, in the recent works of Sir H. De la Beche and Mr. Murchison. The explanations given above was adopted soon after I had finished my Survey of Cumberland, and was published in 1832. See the *Proceedings of the Geological Society of London*, Vol. i. p. 401.

† The brecciated rocks near Barrow, have often been noticed. They are not, however, local phenomena; but belong to the general structure of this middle division of the slates.

filled with agates and other minerals; and they generally effervesce briskly when first plunged in acids.

The position and range of some of the principal slate beds cannot escape notice, as they are often marked by lines of great open quarries. The only difficulty is to know their true dip; for the slaty impress has often destroyed all external traces of the bedding. On this point I must refer to the remarks at the beginning of this letter. One of the best spots for studying, among these old rocks, the difference between *cleavage* and *dip*, is near the jaws of Borrowdale, especially in the great crags which overhang the Skiddaw slate on the north of the gorge.

Of the external features of the lake mountains I attempt not to speak; except so far as they are connected with the inner frame-work of the country. The rocks were elevated and rent asunder—and the rudiments of all the deeper valleys were thus formed in times immeasurably removed from our own days. Again and again have the mountains been shattered by faults, and swept by denuding currents. Their varied structure has produced features of many forms. Some have been worn down by the corroding power of time, and are now buried under soil and moorland; others have stood almost unmoved among the buffetings of the elements, and have an aspect now nearly as rugged as that with which they were first lifted from the sea.

Another zone, belonging to the green slate and porphyry formation, appears on the north side of the third and lowest division of the slate rocks; which thus forms a 'mineral axis' with a repetition of the same formations on its opposite sides. (*See the wood-cut*). This zone begins at Berriar, skirts the eastern side of Carrock Fell, rises into High Pike, and is well marked in Binsey Crag: it afterwards gradually thins away, and it disappears near Brigham. In this range it rests on the Skiddaw slate, and is immediately surmounted by the carboniferous limestone, the upper division of the slates not appearing on this side of Cumberland. Compared with the groups above described, it is in a very degenerate form; it contains, however, almost every variety of rock above noticed. In several parts of it the porphyries so abound as almost to exclude all appearance of true slates. Near High Pike it is penetrated by many metallic veins, probably connected with the causes which produce the syenite of Carrock Fell, and the granite of Skiddaw Forest.

LOWER DIVISION OF THE SLATE ROCKS—SKIDDAW SLATE.*

This division (the true position of which was first determined by Mr. J. Otley) is spread over a large area; being bounded

* No. 7, in the wood-cut.

by the rocks of the preceding division, and the carboniferous zone extending from the old red sandstone, near the foot of Ulls-water, to Egremont. For a few miles south of Egremont, the western end of the Skiddaw slate is immediateiy overlaid by the new red sandstone.

It is of great but unknown thickness; and it has little constancy in its strike and dip, being thrown into great undulations, indicated by the irregular features and varied outline of the country. The coombs and peaks surrounding Skiddaw Forest, and the beautiful succession of grassy mountains between Derwent Water and Crummock Water present the best features of this formation. It is chiefly composed of a dark-coloured glossy slate, occasionally penetrated by great veins of white quartz; and small veins of that mineral are sometimes seen to ramify through every part of the rock; but it contains no organic remains, and hardly a trace of carbonate of lime. Roofing slate has in a few places been obtained from it; but most of the quarries have been abandoned. Occasionally, it passes into the state of a micaceous flagstone, and it alternates, rarely, with coarse gritty beds. On the whole, it is distinguished from the higher groups by its dark colour and fine texture, by the absence of alternating bands of igneous rock, and by its seldom effervescing with acids. Many of the beds of the middle division of slates contain a considerable portion of carbonate of lime and effervesce briskly in acids. Again, in the Skiddaw slate many of the masses flake off parallel to the beds, and the cleavage planes are not so well defined as they are among the green slates; in other places, however the stratification is very obscure. Except as being the base of the whole series of the Cumbrian deposits, and the matrix of some curious metallic veins, this division possesses little comparative interest.

Before I end this sketch of the Cumbrian rocks, I must notice a beautiful group of crystalline slates, which are seen in Skiddaw Forest, between the black slates above described and the granite of the Caldew. If we descend from the high peaks of Skiddaw or Saddleback to any of the bosses of granite which break out near the banks of the rivulet, we cross a series of slaty rocks nearly in the following order:—

1. Dark glossy slate studded with a few crystals of chiastolite. It is overlaid by, and passes into, common Skiddaw slate.

2. A similar slate with more numerous crystals of chiastolite; passing at its lower limit into a hard, shining, sonorous rock, almost made up of matted crystals of that mineral.

3. Mica slate spotted with ill-formed crystals of chiastolite.

4. Quartzose, and micaceous slates of very irregular structure; sometimes passing, when close to the granite, into the form of gneiss.

I believe that this beautiful mineral group is nothing more than the Skiddaw slate, altered and mineralized by the long continued action of subterranean heat. The granite, though a fused rock, may not have produced the whole of this change; but it is at least an indication of the kind of power by which the 'metamorphic' structure was brought about.

I here bring to an end my notice of the Cumbrian slates. To one who is not interested by the complicated structure of the older rocks, it may have appeared tedious and repulsive; but I knew not how to make it shorter, and it relates perhaps to the most difficult chapter in geology.

GRANITE, SYENITE, PORPHYRY, PORPHYRITIC DYKES AND OTHER IGNEOUS ROCKS.

It remains for me to notice a series of rocks, not formed, bed upon bed, by the agency of water; but protruded by subterranean fires among the deposits above described.

Granite of Skiddaw Forest, &c.—I mention this rock first, because it rises out from beneath the oldest strata of Cumberland (*No. 8, in the wood-cut*); and appears to indicate the cause that first elevated the cluster of mountains, of which the peaks of Skiddaw and Saddleback form the highest points. But I can offer no proof that it is older than the beautiful syenite of Carrock, or the granite of Eskdale, or the red syenite of Ennerdale and Buttermere. It breaks out at Syning Gill, between Saddleback and Skiddaw; afterwards at a lower level, near the Caldew, in the channel of which it may be seen for more than a mile; and lastly, about a mile above Swinside, near the first ramifications of the rivulet. At this last place it derives great interest from its near approach to the syenite of Carrock Fell, from its changes of structure, from the mineral veins* by which it is traversed, and from the highly crystalized and altered form of all the neighbouring slate rocks.

Syenite of Carrock Fell, &c.—This beautiful rock exhibits almost endless varieties of structure: but it is chiefly noted for its crystals of hyperthene, and for the great quantities of titaniferous oxide of iron disseminated through its mass. On the eastern side of the hill it passes into a common syenite. In its

* None of the veins were worked to profit when I last visited the spot, nearly twenty years since. They were, however, occasionally opened by mineral dealers: for they contain apatite, schoral, tungsten, wolfram, and several other minerals in considerable abundance. I was struck with the close resemblance of the mineralized portion of Skiddaw Forest to certain parts of Cornwall near the junction of the granite and slate. The physical phenomena are nearly the same; but the Cornish slates are of a much more recent date than the slates of Skiddaw.

farther range towards the east, it becomes almost as compact as basalt, and has, here and there, a globular structure: and, lastly, in its prolongation in the form of a narrow tongue into the extreme branches of the gills on the east side of High Pike, it passes into a felspar rock. This whole mass plunges under a group of igneous and altered rocks: and when on the spot, I considered it only as an instance of one of the porphyries, near the base of the middle division of the slates (green slate and porphyry), in an unusual state of crystallization. Should this opinion (thrown out as a conjecture) be confirmed, we must then consider this syenite as older than the neighbouring granite; for all the granites in the lake country are unquestionably of more recent date than the two lower divisions of the Cumbrian slates.

Porphyry of St. John's Vale—Of this rock (which never, I believe, passes into a true granite) but might be described as a variety of syenite) there are two principal masses—one, stretching for about a mile northwards from St. John's Chapel—the other, of still larger dimensions, ranging in the same direction, on the other side of the valley, from the base of Wanthwaite Crag. Two other small masses break through the Skiddaw slate a little farther towards the east, near White Pike. What was the exact date of the eruption of the plutonic rocks, I do not pretend to determine. When the largest mass was protruded, it bore upon its surface an enormous fragment of Skiddaw slate, which was thus elevated far above its natural level, mineralized by heat, and jammed against the base of Wanthwaite Crag. I mention these phenomena, because they are of great interest to any one who wishes to mark the effects produced by the protrusion of igneous rocks.

The subterranean forces had strength to raise the great masses of porphyry through the soft and yielding Skiddaw slate; but not to push them through the higher group of green slates, which were held together too firmly by the older bands of bedded porphyry to be penetrated by such a movement. Hence it is, that the pophyry of St. John's Vale abuts against, but does not pierce, the middle division of the slates, which range through Great Dod and Helvellyn.—The great 'fault' represented by the deep valley between Raise Gap and the bottom of St. John's Vale, must obviously have been formed after the eruption of the porphyry.

Granite of Eskdale, &c.—This is, out of all comparison, the largest mass of Cumberland granite. It ranges southward as far as Bootle, on the north side of which place it abuts against some highly mineralized Skiddaw slate; and it forms the rugged hills on both sides of the Esk and the Mite, ranging up to the higher forks of those rivers. At its north-western and north-eastern extremities it runs out into two long projecting masses—one of which strikes over Irton Fell and blends itself with the syenite of Wastdale Foot: the other, after ranging along the side of Scawfell, above Burntmoor Tarn, breaks out, here and there, from under the turf-bogs, and passes over the hills into Wastdale Head.

It would be in vain for me, in this short summary, to attempt any regular description of this granite; but the following facts deserve notice:—

About half a mile from Bootle, the granite has been injected, in the form of large ramifying veins, into a black porphyritic rock, which is, I believe, only an altered condition of Skiddaw slate.

In one of the water-courses, in the same neighbourhood, the greater part of the rock is quite earthy in structure; but shows a number of hard spheroidal central masses, like the hard balls in decomposing basalt.

Descending into Wastdale Head by Burntmoor Tarn, we meet with traces of granite veins, and fragments of slate entangled in the granite.

In the upper parts of Eskdale, the granite, in one or two places passes into a nearly compact rock, and has a semi-columnar structure.

At the upper surface of the granite, and near the lines of demarcation between the granite and the slates, there is not unusually a zone of felspathic or syenitic rock, which forms such a passage between the two formations that it is no easy matter to determine the exact boundary line of either. These appearances seem to have been caused by the gradual fusion and altered structure of the masses at the base of the green slate and porphyry.

Red felspathic veins (in structure like the peculiar rocks just noticed) shoot from the granite into the green slate and porphyries. Many examples of this kind are seen in the hills near Eskdale Head.

On the north-western side of Devock Water are many fine masses of crystalline quartz rock close to the junction of the granite and green slate.

Pyritous veins with micaceous iron ore are found here and there, at the junction of the granite and the slate.—Facts like these may help the observer in drawing right conclusions from the intricate phenomena presented by this part of Cumberland.

Syenite of Ennerdale and Buttermere.—This beautiful rock ranges from the neighbourhood of Nether Wastdale Chapel to a point about two miles above the foot of the lake. After being covered by some highly crystalized and rugged masses of slate and porphyry, it breaks out again in Bolton wood, and extends towards the north as far as the side of Reveling Pike; and thence across Ennerdale Water to the Scaw and Herdhouse—at the latter mountain abutting against the Siddaw slate. Its eastern boundary ranges on the north side of Seatallan and the Haycock; and then descends in a long undulating line through the great coves: and crosses the Ennerdale river under the Pillar. The red syenite forms the rugged hills, from the lower part of Ennerdale Water to a point more than two miles above the head of the lake; then ascends towards the N. E. by the shoulder of Red Pike, and thence it may be followed to Buttermere and the hills beyond Scale Force.

After many a toilsome walk, I made out the boundaries of the Eskdale granite and the Ennerdale syenite. But there was no good physical map on which I could lay down my observations correctly. What is here stated may be enough, and perhaps more than enough, for the readers of these letters. The following are the best places for studying the nature of the syenite and its effects upon the stratified rocks:—The junction between the south side of Reveling Pike and the western shore of Ennerdale Water.—The junction of the syenite and Skiddaw slate at Herdhouse.—The south side of the whole pass from Ennerdale by Floutern Tarn to Buttermere; and the whole escarpement under Red Pike, High Stile, and High Crag.—The junctions in the upper part of Ennerdale below the Pillar.

The syenite abuts against the Skiddaw slate at Reveling Pike; and below the junction, in the hills skirting the west side of Ennerdale Water, the slate rocks are much mineralized. Similar effects may be seen on the north side of Herdhouse; where the black slates are so changed that they can hardly be distinguished from the porphyries of the middle division.—Between the foot of Buttermere and Floutern Tarn the phenomena along the line of junction are most varied and instructive The syenite runs through the Skiddaw slate in the form of enormous dykes, or ramifies through it in veins. In some places the formations are in almost inextricable confusion—the slate rocks in one place abutting on the syenite, in another supporting it, and in a third resting upon it.—A great mass of the Skiddaw slate has been caught up by the syenite, carried to the top of Red Pike, and wedged against the green porphyries of High Stile.—Three masses of syenite break through the mineralized Skiddaw slate in the brows overhanging Buttermere; and close to one of them is a mineral vein.—Lastly, where the line of junction crosses Ennerdale, below the Pillar, veins of syenite are seen streaming from the central mass into the green slate and porphyry of the middle group.—In no one case, however, has this syenite in mass penetrated the green slates or passed over them.

Granite of Wasdale Crags, near Shap.—This fine red porphyritic granite is too well known to need description; but the effects it has produced on the neighbouring deposits require a short notice. The rocks on all sides of it are extremely mineralized and changed, apparently by the action of heat. It breaks out at the base of the upper division of the slates, and for some distance appears to have cut off the Coniston limestone. The limestone, however, appears again on the north side of it, and runs down to Shap Wells, but in an altered, shattered, and partly brecciated condition. The flagstones (of the upper division of the slates) are tilted from the granite at a great angle, are much indurated, and have a splintery fracture. Lastly, the slates close to the granite, above Wasdale Head, are completely mineralized, and pierced by small veins of granite injected from the central mass.

As a general conclusion from all the preceding facts, necessarily given in a most condensed form, we may venture to affirm, that all the great masses of porphyry, granite, and syenite above noticed, are rocks of fusion—that portions of them were raised while in a fluid state (otherwise how can we account for the granitic masses injected among the slates);—and lastly, that the same heat which fused the granite or syenite, acting perhaps for many ages upon all the neighbouring rocks, produced that altered and mineralized structure which is so often seen round the centres of eruption.

Porphyritic Dykes, and other Igneous Rocks.—Some of the porphyritic dykes are of great interest; and the subterranean forces by which they were injected among the great breaks and 'faults' of the slate series, have had a very powerful influence upon the position of the beds and the features of the country. A few of them must be noticed.

1. The finest dyke in Cumberland is seen in Kirkfell at Wastdale Head; the mountain has been rent asunder from top to bottom, and a great dyke of granitic porphyry has risen through the fissure. Its junction with the granite at the base of the mountain is not seen, and should it hereafter be found to blend itself with the central mass, it will then be an example of a gigantic granite vein; but from its structure and the straightness of its course, I should rather compare it with the porphyry dykes (or 'elvans') of Cornwall; and if this view be right, it must have been injected through a fissure cutting both through the granite and the green slates. I may here also notice one or two vertical syenitic dykes which rise from Wastdale Head, and cut through the mineralized slates between Great End and Scawfell Pikes.

2. A beautiful dyke of red syenitic porphyry may be traced from the crown of the hill west of Thirlmere into a great water-

course above Armboth. It shows many changes of structure, and is in some places almost compact at its junction with the slate; in which respect it is similar to many Cornish 'elvans.'

3. Many striking examples of red porphyritic dykes are seen in the channel of the Duddon below Seathwaite, and in the hills on the west side of the river. They are seen also on the north side of Black Coomb, and in one of the deep gills that descends from the north-eastern side towards Bootle; and on its south-eastern side granitic dykes break out near its base. Black Coomb is of contorted Skiddaw slate; and has by a great 'fault' been raised two or three thousand feet above its natural level. May we not conclude, that the same subterranean forces which rent the solid rocks asunder and poured the dykes of molten matter through the cracks, employed also their strength in dissevering whole mountains, and elevating Black Coomb into its present position among the green slates and bedded porphyries?*

4. There are five places, not far from the Shap granite, where red porphyritic dykes come to the surface—on the north side of Wet Sleddale—in the valley above High Borough Bridge (the dyke strikes nearly north and south and descends towards Bannisdale)—on the crown of the hill at the right hand-side of the road ascending from the same place towards Shap—and in two places farther north, and near the road side. These dykes cannot, I think, be properly described as granite veins; because no veins resembling them are seen near the junction of the granite and the slates. They are, however, indications of the same powers of nature which produced the granite, but acting at a later period.

5. Lastly, to avoid details inevitably dry and tedious, I may add, that dykes resembling those above described are found near the foot of Coniston Lake—on the road between Coniston and Hawkshead—on the north side of Middleton Fell—and among the slate rocks between the valleys of Dent and Sedbergh.

All the preceding dykes were, I believe, injected before the period of the old red sandstone. But there are, among the Cumbrian mountains, masses and dykes of dark-coloured trappean rock, sometimes approaching the structure of basalt, which are

* Any one who takes an interest in these phenomena, would do well to make a traverse from the south-western shore of the Duddon sands to the Whicham valley, and thence over Black Coomb to Bootle. On this line the formations appear in the following order:—Mountain limestone (Hodbarrow Point, &c.)—Dark coloured slate and flagstone—Coniston limestone—Green slate and porphyry (Millum Park)—Skiddaw slate, at a low level on the south-east side of the great 'fault.' All the preceding groups dip toward the S. E. The great 'fault' ranges down the Whicham valley, and on the north-western side of it the contorted beds of Black Coomb are brought up with a dip reversed towards the N. W. In the remaining part of the section over Black Coomb, Skiddaw slate is continued; then porphyry and altered Skiddaw slate; and, lastly, granite and granite veins. The two last are seen near Bootle.

perhaps of a newer date. They perform no part, however, which makes them of any importance to my present outline; and geological dates founded on the mineral structure of plutonic rock cannot much be relied upon.*

In whatever way the mountain masses of granite and syenite were protruded, they must have produced enormous derangements among all the slate rocks. Judging, however, from the Black Coomb 'fault,' and from the dykes in the valley of the Duddon, and at Wastdale Head, in Cumberland, I believe that the greatest elevations and contortions of the slates took place after the eruption of the granite and syenite. The subterranean powers, pent in by the cooling of the plutonic rocks, pushed the whole region upward into an irregular dome. The struggle between the expansive forces below and the tension of the rocks above (igneous as well as aqueous) may have been long continued; the whole slate series may have been thrown into great undulations, and set on edge; dyke after dyke may have been injected; and the highest parts of the dome may have been starred by diverging faults, cutting their way indifferently through slates, granites, and syenites. The valleys now diverging from Scawfell represent the directions of these ancient 'faults:' and many other breaks and faults (represented in direction by the other valleys of the lake country) must have been formed during this period of disruption and confusion, *before* the conglomerates of the old red sandstone were spread upon the outskirts of the mountains.

On the northern side of the district described in these letters, many of the valleys descending from the higher mountains are turned aside by the terrace of the carboniferous limestone; and, after running some distance parallel to its 'strike,' escape through it, by fissures of a newer date. But on the south side, the upper division of the slates was fissured by many great north and south 'faults,' which traverse the limestone without being turned aside by it, and must therefore have been produced at some period after it was deposited. Faults of different ages sometimes intersect one another, and afterwards contributed to form one

* For the sake of those who may wish to examine the country in detail, I may mention a few examples of such dykes as are alluded to in the text:—

The road-side near Long Close, and thence up to the brow of Great Dod, on the eastern side of Skiddaw: in this brow there are many dykes.—The western side of Bassenthwaite Lake.—Near the foot of the same lake, and along the ridge of the hill on the north side of the road from thence to Cockermouth.—Two or three places to the S. and S. E. of Cockermouth.—The left side of the road from Penruddock and Threlkeld, near Lane-head.—These are all in the Skiddaw slate. (In the middle division of the slates there may be many recent trappean rocks: but it must be very difficult to separate them from the old bedded porphyries).—Near Bowland Bridge, on the old road from Kendal to Ulverstone, &c,. &c.

valley. Thus, Langdale and the upper part of Windermere show the direction of one of the old diverging lines of fault: but the lower part of the lake is in the direction of one of the more recent lines of fracture.

In the preceding letters I have endeavoured to explain the structure of the district in the same way in which a mechanist teaches the movements of a machine—by taking it to pieces. All the deposits have been described in a contrary order from that in which they were put together by nature's hand. Let me now endeavour, in imagination, to re-construct the great frame-work of the Cumbrian mountains.

I. Beds of mud and sand were deposited in an ancient sea apparently without the calcareous matter necessary to the life of shells and corals, and without any traces of organic forms.—These were the elements of the Skiddaw slate.

II. Plutonic rocks were then, for many ages, poured out among the aqueous sediments—beds were broken up and re-cemented—plutonic silt and other materials in the finest comminution were deposited along with the igneous rocks—the effects were again and again repeated, till a deep sea was filled with a formation many thousand feet in thickness.—These were the materials of the middle division of the Cumbrian slates.

III. A period of comparative repose followed. Beds of shells and bands of corals formed upon the more ancient rocks: they were interrupted by beds of sand and mud, and these processes were many times repeated; and thus, in a long succession of ages, were the deposits of the upper slates completed.

IV. Towards the end of the preceding period, mountain masses of plutonic rock were pushed through the older deposits —and after many revolutions, all the divisions of the slate series were elevated and contorted by movements not affecting the newer formations.

V. The conglomerates of the old red sandstone were then spread out, by the beating of an ancient surf, continued for many ages, upon the upheaved and broken edges of the slates.

VI. Again occurred a period of comparative repose; the coral reefs of the mountain limestone, and the whole carboniferous series, were formed; but not without many great oscillations between the levels of land and sea.

VII. An age of disruption and violence succeeded, marked by the discordant position of the rocks, and by the conglomerates under the new red sandstone. At the beginning of that time was formed the great north and south 'Craven fault,' which rent off the eastern calcareous mountains from the older slates; and soon afterwards, the great 'Pennine fault,' ranging from the foot of Stainmoor to the coast of Northumberland, and lifting up the terrace of Cross Fell above the plain of the Eden. Some of the north and south fissures (shown by the directions of the valleys leading into Morecambe Bay) may have been formed about the same time;—others must have taken place at later periods.*

VIII. Afterwards ensued the more tranquil period of the new red sandstone; but here our records, on the skirts of the lake mountains, fail us, and we have to seek them in other countries

IX. Thousands of ages rolled away during the secondary and tertiary epochs. Of those times we have no monuments in Cumberland. But the powers of nature are never in repose; her work never stands still. Many a fissure may in those days have started into an open chasm, and many a valley been scooped out upon the lines of 'fault.'

X. Close to the historic time, we have proofs of new disruption and violence, and of vast changes of level between land and sea. Ancient valleys may have been opened out anew, and fresh valleys formed by such great movements in the oceanic level. Whatever strain there may have been in the more solid parts of our island at this time, their greatest power must have been exerted upon ancient valleys, where the continuity of the beds was already broken. Cracks among the strata may, during this period, have passed into open fissures—vertical escarpments have been formed by unequal elevations on the sides of the lines of fault—and subsidences have given rise to many tarns and lakes. The face of nature may therefore have been greatly changed while the land was settling to its present level.

But let me not be misunderstood; this last period may have been of very long duration. I am only attempting to give an outline of a long series of physical facts, proved by physical evidence. I wish to pause before I reach the modern period; and do not profess to link geology to the traditions of the human race. By some rash and premature attempts of this kind, much harm has been already done to the cause of truth and Christian charity. While geology is an advancing science, and the limits of her discoveries are so ill-defined, such attempts must almost inevitably involve some of the elements of error, and end in uncertain conclusions, ill fitted to form the base of historic truth.

Any description of the mineral veins of Cumberland would involve me in difficult details quite unfit for these letters; and with their present condition, I am not acquainted.—The anti-

* The magnesian conglomerates near Kirkby Stephen rest, almost horizontally, on the beds set on edge by the "Craven fault." But near Brough the same conglomerates are set on edge by the "Pennine fault." Hence we infer that the "Craven fault" was of an earlier date than the "Pennine."

mony works in the Skiddaw slate, near the foot of Bassenthwaite Lake, are, as I am informed, now deserted.—Ores of lead and copper are still extracted from several parts of the middle division of the slates. The large works near Ullswater and Coniston Water Head well deserve a visit.—The mines of *plumbago*, or black lead (*carburet* of iron), near the head of Borrowdale, are so peculiar to Cumberland that they must not be entirely passed over. The mineral is found in a large and very irregular vein, cutting through the green slate and porphyry—not in ribs parallel to the sides of the vein, nor in the form of crystalline masses imbedded in spar; but, here and there, in large irregular lumps, or a congeries of lumps, which begin and swell out, and then thin off, without any apparent order. The miners have sometimes followed the vein for years without stumbling on any of the larger rich masses,* and the works are now, I believe, very unproductive. Several irregular veins, with much red oxide of iron, are found in the neighbouring hills; but none of them have produced the lumps of carburet of iron.—Plumbago is sometimes found in small flakes among the slags of our great iron furnaces; and it has also been found among coal strata near the sides of 'trap dykes.' In such cases we can give an intelligible account of its formation: but I do not venture to account for its sublimation among the rocks of Borrowdale. I may, however, observe that the Skiddaw slate, which supports the green slate and porphyry, sometimes, I believe, contains a small quantity of carbon.†

The iron mines of Low Furness, and of Bigrigg Moor, near Whitehaven, are also characteristic of the lake country. Red oxide of iron has been produced abundantly during many geological periods; and the old red sandstone derives its colouring matter chiefly from that mineral. But the great deposits of 'kidney ore,' near Dalton and Whitehaven, are of a newer date; as they are found in the fissures and hollows of the carboniferous limestone. They in some places mark the presence of a great irregular 'fault;' in others they have been precipitated in open water-worn caverns. The best example of the kind is seen at Bigrigg Moor.—In all these places the 'kidney ore' was probably introduced during the period of the new red sandstone, while the waters of the sea, saturated with red oxide of iron,

* One of the largest masses ever found in this mine, yielding about 70,000 lbs. of the purer sorts of this mineral, besides more of an inferior quality, was discovered about forty years ago.

† A sub-carburet of iron is found in very thin veins or 'strings,' among the slate rocks of Cornwall, north of the Lizard district. But there the slates are perhaps not older than the lower part of the old red sandstone; and I may remark that carbonaceous matter and many impressions of plants occur in the Rhenish provinces, among still older rocks: but among none of such antiquity as the Skiddaw slate.

flowed through the fissures and caverns of limestone, and filled them gradually up with the metallic matter held in partial solution.

In ending this imperfect outline of the structure of your native mountains, permit me to add one or two remarks, not, I trust, unconnected with the object of these letters. Geology links itself with every material science. The earth is a great laboratory and storehouse of old experiments, wherein we may discipline our thoughts, and rise to the comprehension of the laws of nature: and it is by such means that we learn to bring the materials around us under our control, and make them obedient to our will. Exact science is the creature of the human mind—a body of necessary truths built upon mere abstractions. But when physical phenomena are well defined, and their laws made out by long and patient observations, or proved by adequate experiments, they then, by an act of thought, may be made to pass into the form of mere abstractions, and so come within the reach of exact mathematical analysis: and many new physical truths, unapproachable in any other way, and far removed from direct observation, may thus be brought to light, and fixed as firmly as are the truths of pure geometry.

Laws of atomic action—all that belongs to the highest generalizations of chemical philosophy, may gain light and strength from the advances of geology. For what are crystalline rocks, and cleavage planes of slates, and all the perplexing phenomena of metallic veins, but the results of chemical action carried on upon a gigantic scale—of experiments made of old in nature's laboratory—which we can sometimes feebly imitate? The laws of electro-chemical action are among the great discoveries of modern times. We can now separate metals from the fluid in which they are dissolved, in imitation of what nature has done among the cracks and veins of our ancient strata. It is not possible to tell what great things may not hereafter be brought to pass by this happy union of observation and experiment.

Again, we are assured from direct observation, that the same chemical and mechanical laws by which the materials of our globe are now bound together, have remained unchanged from the time when the solid foundations of the earth were laid. Changes of phenomena imply only a change of conditions, not a change in the primary laws of matter. We may therefore hope that, as geology advances farther towards exactness as a science of observation, its phenomena may be brought more nearly under the government of known mechanical laws, and more closely defined by the powers of exact calculation. For ages to come, geology may offer problems to call forth the utmost skill of mechanical philosophy. The density of the earth's mass is not yet exactly known; and no one perhaps has yet found where he is to fix the fulcrum of the lever which is to weigh the world. I believe that this problem will one day be more exactly solved (as it was a few years since attempted) by observations at the bottom of a mine; where geology and astronomy, aided by the refinements of mechanical skill, must all combine in a common labour. This object, if once gained, would not be sterile; but would be pregnant with many results of deep physical importance. But it would be idle for me to dwell on the prospects of geology, or on its bearings on the progress of the exacter sciences. Let me, however, add, that as all parts of nature, material and moral, are the offspring of one Creative Mind, and are wisely fitted to one another; so we believe that the discovery of every new physical truth must tend to the support of every other truth, whatever be its kind, and to the good of the human race.

The great formations of geology, however varied in their features, or imposing in their combination, derive their chief interest from being the monuments of successive periods of time. There is, therefore, a kind of historical animation in our labours which hardly belongs to any other physical pursuit.—The same remark applies to the organic remains buried among the successive strata of the earth. However instructive they may be, in showing us certain forms of organic life, and whatever delight they may give the naturalist, by enabling him to fill up great chasms in the history of animated nature; in the mind of the geologist they have a still higher value, when he regards them as the marks of Creative Power which called into existence successive races of beings adapted to successive conditions of the earth. In this view, they have been not unaptly called 'the medals of creation'—each series marking but one chapter in the physical records of past time.

There is one view of geology, considered by some as a sign of its imperfection, but which, in truth, is a part of its glory. Many of its conclusions are as firmly fixed as the truths of demonstration; but the boundaries of its conquests are still undefined; and there is still so much of wild untamed nature about it, that it is almost as well fitted to inflame the imagination, as to inform the reason. We profess to build only on observation and experiment: but there are many wide provinces in geology still unexplored; many that are known imperfectly; and in no part of her realms are her subjects bound by such unyielding fetters as to have no room for the mind's creative powers. While we are moving on towards a resting-place we are longing for, among objects which to many may seem harsh-featured and repulsive, we may refresh our souls by sometimes soaring into the airy regions of hypothesis, or in fostering dreams as wild as those of a poet's fancy.

You, Sir, have told us of 'the mighty voice of the mountains,' and have interpreted its language, and made it the delight of thousands: and, in ages yet unborn, the same voice will cheer the kindly aspirations of the heart, and minister to the exaltation of our better nature. But there is another 'mighty voice,' muttered in the dark recesses of the earth: not like the dismal sounds of the Lebadean cave; but the voice of wisdom, of inspiration, and of gladness; telling us of things unseen by vulgar eyes—of the mysteries of creation—of the records of God's will in countless ages before man's being—of a Spirit breathing over matter before a living soul was placed within it—of laws as unchangeable as the oracles of nature—of harmonies then in preparation; but far nobler now that they are the ministers of thought and the instruments of intellectual joy; and to have their full consummation only in the end of time, when all the bonds of matter shall be cast away, and there shall begin the reign of knowledge and universal love.

Whatever be the value of geology as a science, its bearings upon the ordinary wants of life are too obvious to call for any comment. It leads us to the most glorious portions of the world, and carries us amongst men of kind hearts, and upright independent thoughts. It is among the mountains, as you have told us, that we are to listen to 'liberty's chosen music:' and the very objects with which we have there to struggle, give back to us, as the earth's touch did of old to the giant's body, new spirits and enduring strength.

Some of the happiest summers of my life were passed among the Cumbrian mountains, and some of the brightest days of those summers were spent in your society and guidance. Since then, alas, twenty years have rolled away: but I trust that many years of intellectual health may still be granted you; and that you may continue to throw your gleams of light through the mazes of human thought—to weave the brightest wreaths of poetic fancy—and to teach your fellow-men the pleasant ways of truth and goodness, of nature and pure feeling.——But here I must conclude my letters; which though of more than twice the length I first intended, do not contain a hundredth part of what might be said on the structure of your country. Such as they are, I send them to you with great good-will; and rejoice in the thought of having at length performed a promise, made to you many years since, but claimed by you only now. With the honest expressions of admiration and regard, and with hearty wishes for your happiness, I remain, &c.

A. SEDGWICK.

Cambridge, May 30, 1842.

SUPPLEMENTAL LETTERS.

LETTER I.

My dear Sir,—Since the three preceding letters were written, in 1842, I have twice visited the mountains of High and Low Furness, the most interesting portions of the country between the Coniston limestone and the banks of the Lune, and the whole range of mountains on the east side of the Lune, between Ravenstonedale and Kirkby Lonsdale. My observations on these tracts of country have been embodied in two papers published in the Quarterly Journal of the Geological Society;* to which I may refer for numerous sections and many details unsuited to my present letter. The whole series of slate rocks which extend from the centre of Skiddaw Forest to the banks of the Lune, may be separated (as was first shewn by Mr. Otley) into three great subdivisions: viz.

I. Skiddaw slate.

II. Green roofing slate and porphry.

III. Dark-coloured slate and flagstone, alternating with bands, and sometimes with thick beds, of siliceous gritstone.

These three primary divisions have been described in my former letters in descending order; and are delineated on the left side of the first wood-cut, where No. 8 represents the central granite; No. 7, the Skiddaw slate; No. 6, the Green slate and porphry; No. 5, the Dark slates, &c. resting on their band of Coniston limestone; No, 4, the Old red sandstone; No. 3, the Carboniferous limestone. This is a true geological sequence: but no attempt is made, in this small wood-cut, to represent the flexures of the beds or to convey any notion of the features of the country.

The Skiddaw slate I am not able to bring into close comparison with any of the great rock formations of North Wales; but the green slate and porphry of Cumberland are so identical in position and structure with the corresponding rocks of N. Wales (expanded through Carnarvonshire and Merionethshire, from the Menai Straits to the Berwyn chain), that I refer them all to one geological epoch. There is however, one marked difference between these great deposits of Wales and Cumberland. The former contain several bands of fossils, but none have yet been

* Vol. i. p. 442.—Vol. ii. p. 106.

found in the latter. In my third letter, I have endeavoured to explain the cause of this difference, and I cannot now dwell on matters of speculation. But there remains another question. *Into how many natural groups may we separate the third and highest division of the slate rocks of the Lake District?* A very inadequate answer was given, in my third letter, to this question. The boundary line between the upper and lower Silurian rocks was drawn hypothetically in 1842, when I estimated the lower Silurian rocks at a much greater thickness than they proved to have been on closer examination. The consequence was that I arranged the *Ireleth slates* (I now think, erroneously) in the same group with the *Coniston limestone*, and placed them both nearly on a parallel with the *Caradock sandstone* of Sir R. I. Murchison.

To help the reader to comprehend descriptions, of necessity short and imperfect, I subjoin a second wood-cut (No. 2), in which the groups are arranged on an ideal section in the *ascending order*. This section is supposed to begin with the rugged mountains near the head of Windermere or of Coniston Lake—thence to be carried southwards to the valley of the Kent, near Kendal—and from the valley of the Kent to that of the Lune above Kirkby Lonsdale. No attempt is here made to delineate any of the breaks and flexures of the beds. Those who wish for more exact and detailed sections may consult the Quarterly Journal of the Geological Society (Vol. ii. pp. 106—131).

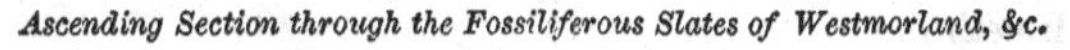
Ascending Section through the Fossiliferous Slates of Westmorland, &c.

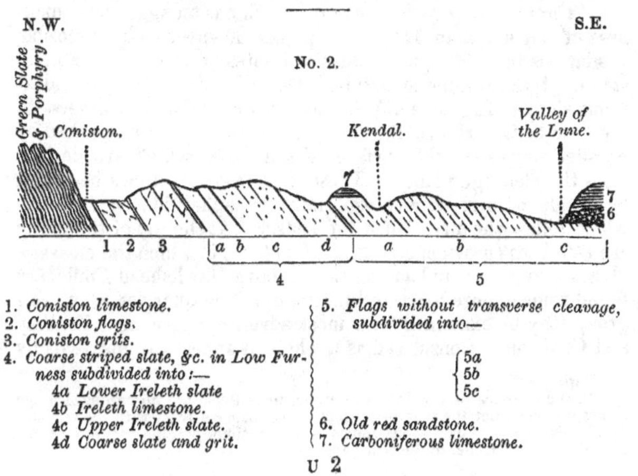

1. *Coniston limestone.*
2. *Coniston flags.*
3. *Coniston grits.*
4. *Coarse striped slate, &c. in Low Furness subdivided into:—*
   - 4a *Lower Ireleth slate*
   - 4b *Ireleth limestone.*
   - 4c *Upper Ireleth slate.*
   - 4d *Coarse slate and grit.*
5. *Flags without transverse cleavage, subdivided into—*
   - 5a
   - 5b
   - 5c
6. *Old red sandstone.*
7. *Carboniferous limestone.*

1. *Coniston limestone and calcareous slate, &c.*—This group is too well known to need any detailed description. The more pure calcareous bands are generally found in the lower part of it, are seldom of any great thickness, and often become almost entirely degenerate and worthless. The upper part of the group consists of dark-coloured, pyritous, earthy slates and shales, seldom sufficiently indurated to be of any value. In its whole range* from Shap Wells to Duddon Bridge its average thickness is not more than 300 feet; and its fossils, especially abundant in the calcareous shales, are lower Silurian (see Appendix B). At Graystone House, the limestone beds are much mineralized by the action of the neighbouring porphry; beyond which place, an enormous dislocation has thrown the range of the whole formation about two miles out of its former bearing. At the S. end of Cumberland, the calcareous bands predominate over the shales, and are worked in some fine open quarries near Beck and Water Blain. The thickness of the whole group is, at this extremity of the range, not less than 600 feet—about double the average thickness of the same group on the other side of the Duddon—and all the beds are highly calcareous; those in the lower part of the group forming a pure dark-coloured limestone with some large white veins. Lastly, I may state that a great flexure or dislocation of the strata has brought up the Coniston limestone at High Haulme, near Dalton, on the south side of the Duddon estuary: but this limestone must not be confounded with a second and higher calcareous band (4*b*), subordinate to the great group of Ireleth slates.

2. *Coniston flags.*—This group, which has an aggregate thickness of not less than 1500 feet, passes downwards by insensible gradations into the upper shales and slates of the Coniston limestone. Its prevailing structure is that of a dark-coloured calcareous slate or flagstone affected by cleavage planes transverse to the true beds. Hence the lines of deposit are often marked by parallel stripes on the surface of the flags, which are derived from the cleavage planes. These lines may sometimes be traced by bands of pyrites, and especially by spheroidal concretions which follow the direction of the beds.† At the top of the group are thin beds (provincially called *sheerbate*) in which the cleavage planes disappear; and among these, above Hawkshead Fould, are found some impure bands of limestone. The quarries where this group may be examined with most advantage are near Brathay and Coniston. Considered as a whole, it appears to me perfectly

* Supra, Note, p. 200.

† In the quarries of Coniston limestone, near Beck, are numerous cleavage planes distinct from the true beds; and there also we meet with the very unusual phenomena of calcareous spherical concretions parallel to the cleavage planes, and not parallel to the true beds.

identical with the lower groups of what I have called Denbigh flagstone. But why should we separate this group from the shales and slates* which rest immediately upon, and seem to group with, the Coniston limestone? The answer is given in the list of fossils. The fossils from the Coniston limestone, &c. form a most characteristic lower Silurian group. But there are no characteristic lower Silurian fossils in the Coniston flags. The fossils are few in number; but they either belong to known upper Silurian types, or to species which have not yet been determined. Among the concretions, so abundant in this group, are impressions of *Graptolites ludensis;* and in the middle and upper portions of it are one or two upper Silurian *Trilobites,* one or two species of *Creseis,* and beds of *Cardiola interrupta.* On the evidence, then, both of mineral structure and of fossils, we are compelled to separate the Coniston flags from the Coniston limestone and calcareous slates, placing the former at the base of the upper Silurian series of the Lake district; and, therefore, nearly on a parallel with the 'Wenlock shale' of Sir R. I. Murchison.

3. *Coniston grits.*—This group is of great thickness, and is composed of such unbending materials, that it exhibits no great flexures or contortions, though its beds are very highly inclined, and sometimes vertical. It is composed of thick beds of grey, or bluish-grey gritstone, alternating, indefinitely, with sandy micaceous shales, and with beds resembling the flagstones of the lower group (No. 2). Both the gritstones and the shales are marked by spheroidal concretions, sometimes of large size. The great gritstone beds are not unusually divided into irregular prismatic masses by joints nearly perpendicular to the planes of stratification, so as to make it difficult to ascertain the true bedding of the rock where the subordinate slates and flags are wanting. Fossils are extremely rare in this group, but some have been discovered in it; and among them *Graptolites ludensis, Cardiola interrupta, Orthoceratites Ibex, O. subundulatus,* and fragments of Trilobites, &c. All the known species are *upper* Silurian. The rocks above described are sufficiently well defined to be laid down upon a geological map; and are geologically important as forming the true base of the great group to which I have given the name of 'Ireleth slates.'†

* See Quarterly Journal of Geol. Soc. Vol. i. pp. 17—20.

† The Coniston grits become degenerate at the N.E. extremity of their range over Shap Fells. They cross the road between High Borrow Bridge and Shap, at the hill top near the Demmings. Farther to the S.E. they are expanded into a well-defined formation, and may be seen ranging, in grey gnarled elevations, by Bannisdale Head to the hills N. of Long Sleddale Chapel—thence in a broad zone to the foot of Kentmere Tarn, and over the bare hills as far as the trigonometrical station N.E. of Elleray. Their strike afterwards carries them over Windermere to Latterbarrow, and thence by the head of Esthwaite Water and Grisedale Head, to the eastern margin of Coniston Water. After crossing the

4. *Ireleth slates.* This group occupies a tract of country, on the average not less than six or seven miles in breadth; and after passing over Windermere into Lancashire, it is spread over a still wider zone. Of its thickness (which is unquestionably very great) we might form a very exaggerated estimate, did we not take into account the continual undulations by which the same beds may be repeated again and again over districts of wide extent. The breaks and contortions—the changes of strike—the anticlinal and synclinal lines, are almost without number. These phenomena meet us wherever we make a long traverse across the beds of this formation. The prevailing rock of the whole group is a rather quartzose slate with transverse cleavage planes which preserve their parallelism even among the most contorted strata. It exhibits the sedimentary *stripes* and other phenomena of cleavage planes in great perfection; but very few of the beds, if we except the Ireleth country, have been found of much value as roofing slate. The more slaty bands alternate indefinitely with quartzose bands, and sometimes with great beds of gritstone, or very coarse greywacké, like those of No. 3. Here, however, the coarser beds are subordinate to those with a slaty impress: while, in No. 3, the coarse grits give the chief impress to the group.—In the country between Coniston Water foot and Shap Fells, I am unable to separate the formation here described into any natural subdivisions or sub-groups; but in its range through Low Furness it may be subdivided as follows:—

(4*a*). *Lower Ireleth slates.*—These beds rest immediately on the Coniston grits (No. 3), are very highly inclined, and occupy a zone more than half a mile in width. They must, therefore, be of considerable thickness.

(4*b*). *Ireleth limestone.*—It breaks out on the brow of a hill about the third of a mile N.E. of the village of Ireleth, and appears to be continuous along the line of strike for about two miles; for it may be traced through five old quarries, one of which (at Meer Beck) is still worked. At first it is of considerable thickness, perhaps twice as thick as the purer bands of Coniston limestone (No. 1), as seen in their range through Westmorland; but it soon degenerates and disappears altogether. After an interruption of several miles, it breaks out again on the S. side of Tottlebank Fell—not in any well-defined beds, but rather in the form of discontinuous calcareous concretions, which may be traced at intervals for more than a mile; and it finally disappears in the hills on the left bank of the brook which de-

water they are thrown (by the same dislocation which has affected the Coniston limestone) nearly a mile out of their former line of bearing, and re-appear in the bare hills S. of Torver Chapel. From these hills they pass, without any apparent deviation, to the low hills S. of Broughton, where they are cut off by the Duddon estuary, and are seen no more on the surface.

scends from Beacon Tarn. A remarkable quadrangular crinoidal stem (*Tetracrinites*) seems to be characteristic of this calcareous band; which also contains innumerable fragments of shells, unfortunately too imperfect to be made out.

(4*c*). *Upper Ireleth slates.*—Among these beds are the great slate quarries of Kirkby Ireleth.*

(4*d*). In this subdivision I include all the remaining rocks between the preceding group (4*c*), and Leven sands, also the slate rocks of Cartmel Fells, and of the contorted ridge which runs from Newby Bridge to Lindal. It is of a coarser and more mechanical structure than the *upper* Ireleth slates (4*c*), and contains hardly any slate beds which are now worked with profit. It is undoubtedly of very great thickness, but a large abatement must be made in our estimate, not only on account of the numerous undulations of the strata, but also on account of the breaks and dislocations by which whole mountain masses have been shifted out of their true line of bearing, so as to destroy the continuity of any one line of section.

The fossils in Low Furness (where the fourth group of the general section admits of the above-named subdivisions) are by no means abundant. In (4*a*) I have found *Graptolites ludensis;* but the deposit is ill exposed, and has few open quarries. The fossils of (4*b*) have been noticed above. In (4*c*) are found *Graptolites ludensis*, a *Cyathophyllum*, *Favosites alveolaris*, and two or three *Orthoceratites*. In (4*d*) occur, though very rarely, corals, encrinite stems, and *Cardiola interrupta*. All the species that have been made out are *Upper Silurian*.

In the range of the beds towards the N.E., the sub-group (4*d*) appears to thin off, and loses all its distinctive characters. The calcareous bands (4*b*) are also wanting. Hence, all the four subordinate groups are packed in one inseparable mass of very great thickness. As, however, the base line of *Coniston grits* (No. 3) is constant, we may approximate to the relative position of the beds above described. For example, the slaty rocks of Bannisdale Head, and those N. of Long Sleddale chapel, appear to be on the exact parallel of (4*a*). The slate quarries of Bretherdale are probably on the parallel of (4*c*). The fossiliferous slates between Underbarrow and Crook are high in the series, and therefore on the parallel of (4*d*). Not, however, to dwell on minute points of comparison, I may state generally—that the whole group I am describing (No. 4) contains several bands of fossils, all of which, so far as the species are known, are *Upper* Silurian. *Terebratula navicula* (along with eight or ten Upper Silurian species) occurs near the base of the group above the

* For some details respecting these quarries which cannot be given here, I may refer to the Quarterly Journal of the Geological Society, vol. ii. p. 114, &c.

Ferry House on the shore of Windermere; and also occurs in great abundance in the upper beds between Crook and Underbarrow.*

On a review of all the facts above stated, I now think it certain that the *Coniston flags* and *grits*, and the *Ireleth slates*, cannot be classed with the *Coniston limestone*, and that, considered collectively, they are very nearly the equivalents of the *Wenlock shale* and the *lower Ludlow rock* of the Silurian System.

The southern and upper boundary of the great group above described (No. 4) would be approximately defined by a curve line drawn near the base of the hills between Lindale and Crosthwaite, and thence to Underbarrow chapel. From the chapel, it might be drawn along the valley to Brundrigg; thence, on the N. side of Burneside mills, across the Sprint to Garth Row. From the last named place it might be drawn nearly E. and W. as far as the valley of the upper Lune. To the S. of this line is a remarkable group, generally ill exposed, but occupying a country about two miles in width. The lower part of it should be classed with No 4; the upper part with No. 5 of the ascending section. But, from the obscurity of the country, no continuous demarcation can be traced on a geological map. To this last remark the natural section between Underbarrow chapel and the limestone of Kendal Fell forms so remarkable an exception, that it deserves a short notice.

Leaving then the hills which extend, with many undulations, from Mountjoy to Crook (about the age of which there can be no doubt, for they contain *Terebratula navicula* and several other characteristic fossils, and have, in perfection, the peculiar structure of the Ireleth slates, No. 4), and commencing an ascending section with the faulted beds of the valley near Underbarrow chapel, we find, on the line of the Kendal road, the following series well exposed, and not interrupted by any contortions of the beds:†

(*a*). A thick group of coarse slate and flagstone, with a rude and very imperfect cleavage, extending nearly to a farm called High Thorns. In mineral structure this group more resembles No. 5 than No. 4 of the wood-cut; but it contains *Terebratula navicula*, and may, perhaps, be considered as the uppermost limit of No. 4, making a passage into No. 5.

(*b*). A bed six or seven feet thick, with two species of *Asterias*, one, *A. primæva* (Forbes' M.S.), the other new. Here the

* See Geological Society's Journal, Vol. ii. p. 123.

† The length of this section measured directly across the strike is more than a mile, and all the beds dip at a high angle towards the East. The whole thickness of the beds between Underbarrow and Kendal Fell must therefore be very considerable, and might very easily be measured.

Cyathophyllum disappears on the asscending section, and the *Terrebratula navicula*, one of the most characteristic fossils of the great group (No. 4) is no longer found.

(*c*). Flags, *sheerbate* (i. e. without cleavage); some red calcareous bands with many fossils. Numerous *Trilobites*, one like *Calymene Blumenbachii*.

(*d*). Striated hard grits and *sheerbate* flags, with many fossils identical with those under Benson Knot, on the S. side of the Kent. The Asterias beds are found at Docker Park, under Benson Knot; in the valley above Kendal, near Redman Tenement; and also in the river Sprint, about a mile below the Tenement. The upper groups, (*c*) and (*d*) of this section, range, under the great limestone escarpment of Kendal, to Brigsteer: thence they cross the mosses, pass under the 'great scar limestone' of Whitbarrow, and re-appear on the other side at Fell-end. From this neighbourhood they may be followed down to Ulpha Crags, on Milnthorpe sands, where they are finally cut off by the sea.

5. *Upper slates of Kendal and Kirkby Moors, &c.* This great formation may be conveniently subdivided into three subordinate groups,—(5*a*), (5*b*), (5*c*).

(5*a*). Includes at least all the upper portion of the Underbarrow section just noticed, and forms a connecting link, through its fossils, and its structure, between No. 4 and No. 5 of the general section.

(5*b*). This great subdivision is bounded to the north by Tenter Fell, Benson Knot, Docker Park, and Lambrigg Fell. Beyond Lambrigg Park, the line (before running nearly E. and W.) deflects towards the S.E, in consequence of a great disturbance which has brought the beds of (5*a*) into a portion of Firbank Fells. To the E. it is bounded by a great fault which ranges down the valley of the Lune. To the W. by the limestone ridges of Kendal Fell and Farlton Knot; and to the S. it is overlaid by the upper sub-group (5*c*). Some of the beds are composed of a grey, bluish-grey, or greenish-grey gritstone, almost as coarse in structure as the Coniston grits (No. 3), but they alternate with, and are often subordinate to thinner beds of grit and masses of coarse slate and flagstone, in which we lose nearly all traces of the cleavage planes which give an impress to the great inferior group, (No. 4, of the wood-cut.) The formation is intersected by many earthy ferruginous decomposing bands, sometimes associated with carbonate of lime and calcareous grit; and these bands are marked with innumerable impressions of fossils. The fossils of this sub-group are eminently characteristic. In it we first meet with three or four species of large *Aviculæ*, numerous *Meristomyæ*, (a new genus of Salter),

Nucula—the *cingulata* of Hisinger, and *Solenocurtus Fisheri* of the former edition of these letters. Three-fourths of all the Upper Silurian fossils of the subjoined list* are found in this single subdivision (5*b*); and it is not only distinguished by numerous characteristic fossils, but by the absence of several species found in the Ireleth slates (No. 4).

(5*c*.) In this group, of small superficial extent, but of considerable thickness, are included the red siliceous flagstones of the Lune, which underlie the old red sandstone of Red Scar and of Barbon Beck foot, above Kirkby Lonsdale, and form the southern limit of the slate rocks on the 2nd wood-cut. Were it not for a wish to bring them into a comparison with the 'tilestones' of the Silurian System, I should not have separated them from the preceding group (5*b*). For their fossils (especially in some red calcareous concretions, near their upper surface, which might be confounded with the 'cornstones' of the old red sandstone) are, perhaps without exception, identical with those which abound in the rocks between Kirkby Moor and Kendal (5*b*). Again, the red colour of these flagstones is by no means constant; for we find among them beds of purple, grey, bluish-grey, and greenish-grey flagstone, all forming a part of one group; and, on the west side of Underley Park, are beds of indurated shale and flagstone, identical in structure with the most ordinary varieties of the inferior group (5*b*). Perhaps the best example of these red flags (or 'tilestones') may be seen at Helme (a couple of miles south of Kendal). They are of considerable thickness, strike nearly N. W. and S. E., and appear to have been let down into their anomalous position by some of the great faults, which, without any exaggeration, may be said to have broken up all the formations between Kendal and the Lune into great disjointed fragments. At the N. end of Helme (Oxenholme), are some calcareous concretions, and a thin band of impure limestone (noticed in my 3rd letter, p. 199), and near the S. end of the hill are some hard, grey, fossiliferous beds, apparently representing some beds of like structure which break out from beneath the red flags on the old Kendal road a little N. of Kirkby Lonsdale.

In conclusion, I may repeat that the fossils derived from all the rocks above described (subordinate to No. 5), form a most characteristic group, on the exact parallel of the *Upper Ludlow rocks* of the Silurian System.

It is obvious, after a single eye-glance, that we cannot immediately connect together the rock formations on the E. & W. sides of the valley of the Lune. The strata on the E. side of the Lune are raised to an elevation, in some places exceeding 2000

* Appendix B.

feet, and are thrown into great undulations which cannot be traced, if we except a small portion of the upper Lune near Borrow Bridge, to the west side of the river.* As, however, I never saw a proof that any portion of the rocks described in the previous section (wood-cut No. 2) had been deposited on the inclined edges of any lower group, I thought it probable that some portions of the upper Ludlow rock (No. 5) might be caught up among the great folds and contortions of Middleton and Howgill Fells. This expectation was disappointed; for after a carefull examination, I have not found a particle of the upper Ludlow series in any of the mountain chains here noticed. The fells of Middleton, Howgill, and Ravenstonedale, are made up exclusively of four well-defined groups—the lowest being Coniston limestone† (No. 1); the highest, Ireleth slate (No. 4), as delineated on one line of section in the following wood-cut; and any other complete line of section through these fells would give the same result.

Rocks of Middleton, Howgill, and Ravenstonedale Fells.

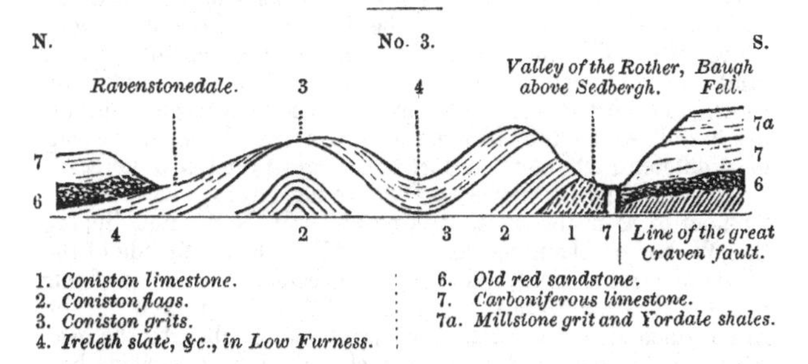

* The connexion between the formations on the two sides of the Lune may, I think, be made on a line of traverse represented, though very inadequately, in a wood-cut of the Journal of the Geological Society (Vol. ii., p. 121, Section 16). It is drawn from the Shap granite to Whinfell Beacon; and thence across the Lune, and over Howgill Fells, to the carboniferous mountains skirting the upper part of the valley of the Rother, above Sedbergh. It may be examined on foot without difficulty in two days, and will well reward the labour. The traverse leads past two noble waterfalls,—one in the upper part of Carnigill—the other, called Cautley Spout, where the water tumbles over scarped edges of the Coniston grits in the precipices which descend toward one of the forks of the Rother.

† Ravenstonedale is formed by a kind of horse-shoe depression among the mountains. Its S. E. end is a prolongation of the valley of Sedbergh, and is drained by the Rother. Its middle portion doubles round the mountains and is drained by a branch of the Eden. Its N. W. end is drained by the Lune. The Coniston limestone runs up the valley of the Rother into the S. E. end of Ravenstonedale. Without this explanation, a part of the subsequent description in the text might seem to involve a contradiction.

The only difficulty is to find a base line on which to construct the sections. That being determined (notwithstanding the vast undulations of the beds, and the great breaks across the chains, accompanied with a change of *strike*), we have little difficulty in making out the successive groups, for they retain their usual structure, and have their characteristic fossils.

The slate rocks, extending from Casterton Fell to the S. E. end of Ravenstonedale, are bounded by the carboniferous limestone, near the base of which runs the great 'Craven fault': and this fault, when it approaches the older rocks, is generally accompanied by a great elevation of the beds on its eastern side. Hence, I was induced to seek, along the line of fault, for the oldest groups of the mountains here described.

Where the Craven fault passes between Ingleton and Thornton Force, a greenish slate rock (overlaid by soft calcareous slates containing irregular bands of impure dark-coloured limestone) is brought up to a considerable elevation, and dips under a great dislocated mass of carboniferous limestone.* I now believe, on analogy, that these calcareous slates represent the Coniston limestone; but they are, unfortunately, so far as I have seen them, without fossils. Whatever may be the age of these calcareous slates, we find nothing like them in the range of the Craven fault on the eastern side of Casterton Low Fell and Middleton Fells. This part of the range cannot indeed give us any continuous base line; for the older rocks strike across the mean bearing of Middleton Fells, and therefore abut against the line of fault.

At the N. end of Middleton Fells we have a magnificent development of Coniston grits, striking about W.N.W. and dipping S.S.W. and overhanging, for several miles, the south side of the valley of Dent. Here, therefore, we reach a low group of our series of deposits. On the west side of the Craven fault, in the ridge which divides the valleys of Dent and Sedbergh, the beds are thrown into a great irregular arch—dipping on one side into the valley of Dent, and on the other into that of Sedbergh. The northern dip is continued into the hills on the northern side of the Sedbergh valley, where the Coniston grits are repeated in their characteristic form. It is obvious, from this description, that we might expect the presence of some of the lowest groups in the ridge above mentioned, between Dent and Sedbergh: and, accordingly, in a deep ravine in Dent, called Helm's Gill, and on the west side of the Craven fault, we find the Coniston group (No. 1), with many characteristic fossils. It is made up of calcareous slate and impure beds of dark-coloured limestone, and is intersected by five or six porphyry dykes. Its

* See Geological Quarterly Journal, Vol. ii., p. 120, Section 12.

relations admit of no doubt; for it is overlaid by, and passes into, a most characteristic group of Coniston flags (No. 2), containing *Graptolites ludensis*, *Cardiola interrupta*, and one or two Upper Silurian *Orthoceratites*. A section commencing with a mountain called Risell, on the north side of the valley of Dent, drawn across Helm's Gill, and prolonged across the valley, and over the north end of Middleton Fells into the valley of the Lune, would give the following sequence:—First, the carboniferous rocks, from the millstone grit to the 'great scar limestone'—cut through by the great 'Craven fault:' and on the western side of the fault, the groups would be—Coniston limestone, &c. (No. 1), Coniston flags (No. 2), Coniston grits (extending across Dent into Middleton Fells) (No. 3), Ireleth slates (No. 4).* Proceeding northwards, the calcareous slates of Helm's Gill disappear under the turf bogs of the mountain ridge, and for several miles we lose all trace of them as we ascend the valley of the Rother above Sedbergh, on the line of the great Craven fault.† Farther up the valley, we meet with calcareous slates with bands of dark limestone, in the gills which descend from Baugh Fell to the Rother. I believe they are continuous for more than four miles; and they finally disappear in the S. E. end of Ravenstonedale, about a mile above Rother Bridge, at a low level, and close to the new road. At this extreme end of their range they contain lower Silurian fossils in the greatest abundance; and they are overlaid, in the brows which overhang the north side of the valley, by the Coniston flags, with *Graptolites ludensis* and *Cardiola interrupta*.

This continuity of the group representing the Coniston limestone (No. 1), is explained by the fact, that the older rocks in this part of the valley, strike in a direction nearly parallel to the range of the 'Craven fault.' The description above given, and the structure of the mountains of Cautley and Ravenstonedale may be explained by help of a section drawn from the N. W. end of Ravenstonedale to Cautley Crags, and from Cautley Crags across the valley of the Rother to the top of Baugh Fell. (Woodcut, No. 3). This section shows the carboniferous rocks, from the millstone grit to the great scar limestone—the Craven fault—and the great undulations of the four groups which begin with the Coniston limestone. The second and third groups are exposed, on a grand scale, in the ascent from the Rother to the head of Cautley crags. The undulations of Nos. 3 and 4 are made out by a traverse down Bowtherdale. The old red sandstone

* See Geological Society's Journal, Vol. ii. page 120, section 13.

† Part of the way along this line, the Conisten group (No. 1.) is probably buried under the conglomerates of the old red sandstone, which are of great thickness.

(placed at the N. end of the section) is not found at Bowtherdale foot; but it does exist at the foot of Langdale, a little further west.*

A general summary of the preceding details contains an answer to the question proposed near the beginning of this letter—"Into how many natural groups may we separate the third and highest division of the slate rocks of the Lake District?"

We have found that it may be separated into *five* natural groups; of which No. 1. represents the *lower Silurian System;* degenerate in thickness, but abounding in characteristic fossils. Nos. 2, 3, and 4, represent the *Wenlock shale* and *lower Ludlow rock*, on a grand scale of development, but with a poor list of fossils. And, lastly, No. 5 represents the *upper Ludlow rock*, on a great scale of development, and with an excellent list of characteristic fossils.—Again, applying the same question to the mountains in the triangular area, bounded on the W. by the great fault of the Lune, on the E. by the great Craven fault, and on the N. by the carboniferous limestone of Ravenstonedale, we find that the fells of Middleton, Howgill, and Ravenstonedale are composed of rocks which may be divided into *four* natural groups, commencing with No. 1, and ending with No. 4, of the general section.

The singular position of these four groups on the E. side of the Lune, and the great expansion of the Ireleth slate group (No. 4) in the country between the foot of Windermere and the sea coast, naturally suggest the hypothesis, that the upper Ludlow rocks of Westmorland (including all the slate rocks between Kendal and Kirkby Lonsdale) were deposited unconformably in a trough or depression of the older strata. But the facts do not appear to bear out the hypothesis. For, so far as I know, there are no examples of conglomerate beds between the groups No. 4 and No. 4; nor have I ever seen the strata of one group resting unconformably on the strata of the other. On the contrary, there are many spots where the strata of No. 5 pass insensibly into the strata of No. 5. Hence we may conclude—that the whole series of slate rocks was deposited conformably, or, at least, without any great mechanical interruption—and that the great disturbing forces, by which the several groups have been so strangely elevated and contorted, did not come into play till near the end of the Upper Ludlow period.

* Some great disturbances which have thrown the calcareous slates, and the beds of Blue Caster, out of their bearing, are not noticed in the text. A little below Rother Bridge are four or five porphyry dykes: and farther down, the rocks are so shattered and contorted that it is, for some distance, almost impossible to define the mean line of strike. These great disturbances appear to have been produced by "cross faults," now indicated by one or two lateral valleys.

Successive periods of elevation, &c. If we make successive traverses through the whole cluster of the Cumbrian mountains (beginning with the rocks described in this letter, and ending with the carboniferous system, which descends towards the Solway Frith and the vale of the Eden), we are at first disposed to refer the elevation of the older strata to the protrusion of the great central masses of syenite and granite. A mechanical protrusion of granitic mountains through one or two central openings, would naturally produce a somewhat circular arrangement of the neighbouring groups of strata: and of this arrangement we do find traces in the rocks round Skiddaw Forest, and in a zone of green slate and porphyry which underlies the carboniferous zone on the northern skirts of the great Cumbrian cluster.* As a much more general rule, however, the older groups of strata strike in a direction not far from N.E. and S.W.: and by help of this direction we may connect the older Cumbrian mountains with those of Carnarvonshire, of the Isle of Man, and of the great chain which forms the southern boundary of Scotland. Hence we may refer the principal elevations of the Cumbrian cluster of mountains to a deep-seated cause, of which the central granite and syenite, above mentioned, are mere local indications. These early movements of elevation took place while the slate-beds were soft and flexible; and the cleavage planes were superinduced afterwards, while the beds were passing into a solid state. We may suppose, on probable evidence, that these movements affected all the groups of strata, including those which now appear on the east side of the Lune. The first contortions of the groups of strata E. of the Lune were undoubtedly effected before the slaty beds were solid, or had cleavage planes. But the great vertical elevation of the mountains—their separation from the neighbouring groups by the great Lune fault—their frequent change of strike—the fragmentary masses into which they have been divided by numerous transverse breaks—these phenomena must belong to later periods. Some of these phenomena may be referred, perhaps, to disturbing forces connected with the protrusion of the Shap granite: but we ought, at the same time, to bear in mind—that the granite has not much deranged the strike of the beds with which it is immediately associated—and that the very complicated phenomena on the E. side of the Lune can hardly be accounted for by any conceivable set of forces emanating from one centre.

However this may be, the chains on the E. side of the Lune had received their present mineral impress before the period of the old red conglomerates: for, if we follow these conglomerates from Kirkby Lonsdale to their last appearance in the valley of

* See wood-cut (No. 1.) and Letter III.

the Rother, about two miles and a half above Sedbergh, we may find, among their rolled fragments, masses of the slate rocks with their cleavage planes, lumps of Coniston limestone with Coniston fossils, and lumps, more or less rounded, of all the other groups, not excluding the 'tilestone'. The appearance of what seemed Coniston limestone among the pebbles of the old red sandstone, above Sedbergh, at one time seemed to me almost inexplicable: but it now offers no difficulty: for the old red sandstone, above Sedbergh, was actually deposited in an ancient oceanic valley partly scooped out of the Coniston limestone and flagstone. Large portions of these conglomerates have been removed by subsequent denudation, and they may never have been perfectly continuous. They are now found sticking, here and there, to the old rocks on the sides of the valley, and sometimes several hundred feet above the level of the river. In this respect they present striking analogies to the *new red conglomerates* in the ancient valleys of the Mendip Hills.

As a general rule, the old red conglomerates are unconformable to all the older rocks of the country here described. But we must bear in mind that among the most contorted groups we may have some beds which are nearly horizontal; and, in such a case, an overlying unconformable deposit may seem, through a short space, to be perfectly conformable to the beds on which it rests. In this way I would explain the phenomena of Red Scar, above Kirkby Lonsdale. There is not, I believe, there any true passage between the tilestone and the red conglomerates; for the red calcareous concretionary beds between the two contain only Ludlow fossils, and do not represent the 'cornstone;' and the red conglomerates contain water-worn fragments of the "tilestone" group. But we do find, here and there, among the conglomerates, calcareous concretions which are of a newer age, and may perhaps represent the "cornstone."

Were it possible to follow out the subject here, I might shew that the great 'Craven fault' was formed before the existence of the new red conglomerates in the valley of the upper Eden: and I might speculate on the epochs of the great faults, ranging nearly north and south, which have broken the carboniferous series into fragments along the south-western skirts of the slate groups above described. During these epochs may have been formed some of the great breaks in the mountains east of the Lune, which are now marked by deep valleys, up which we cannot trace the older red conglomerates.

But none of these old disturbing forces, nor all of them together, can explain the present configuration of the mountains and valleys. Countless ages passed away; and at the dawn of modern times, we have proofs of great changes of level between sea and land, producing successive periods of diluvial drift: and we have proofs of a period of refrigeration, when our higher valleys gave birth to glaciers. By a new change of level, the glaciers were borne away by the sea, and became rafts for the transport of innumerable bowlders from one mountain top to another. Strange as such facts may sound to ears that have not before heard of them, I believe they are capable of perfect physical demonstration. Old as such phenomena may be, in respect to ourselves, they are, in the language of geology but things of yesterday; for I have shewn, in my first letter that the rivers of our present valleys have been flowing through them only a few thousand years: and as for the bowlders which have floated over our valleys, and are, in thousands of places, stranded on the mountain sides, and sometimes perched on their tops—many of them still clink under the hammer, and look as fresh as if they had but just started from their parent seat.

This letter, though addressed to a Poet, is meant chiefly for the eye of those who make geology their study. To many readers, the previous details may seem insufferably technical and repulsive. They were, however, unavoidable. For how could I, otherwise, attempt to grapple with some of the most difficult problems in geology?—or what right had I to dictate opinions without giving some of the evidence on which they were founded?

I subjoin, in an Appendix, a notice of the porphry dykes of the Cumbrian mountains; and an excellent list of fossils, prepared by my friend Mr. J. W. Salter, who is now employed in figuring the species that are new, or which have been imperfectly described before. When the classification of the rocks is firmly settled, and their fossils are figured and described, a geologist may then take the minuter facts for granted, and give the physical history of the country in general and graphic terms, without entering on many of the dry and crabbed details of this letter. Such as it is, my dear Sir, I trust you will receive it with the same good will with which I now offer it, and once more accept my fervent and honest wishes for your happiness and health.

I remain, &c.

A. SEDGWICK.

Cambridge, May 30, 1846.

LETTER II.—(5th of the Series).

TO MR. J. HUDSON, BOOKSELLER, KENDAL.

Dear Sir,—Your request that (with a view to a new edition of your Work) I should make what changes I thought expedient in my four Letters on the Geology of the Lake District, reached me while I was in residence at Norwich, where I had access to none of my previous papers connected with the subject. Had your request reached me at Cambridge, I should perhaps have retained, very nearly in their original form, my three Letters addressed to Mr. Wordsworth in 1842: but I should certainly have re-written my fourth Letter, that was published in the spring of 1846. In consequence of this unavoidable delay in my reply, I now learn that my four Letters will appear, word for word, as they were printed in your former edition. I hardly regard this as any misfortune: for the historical progress of any work of scientific arrangement is sometimes highly instructive, while it shows us the difficulties that retarded the progress of the Work, and the way in which they were gradually overcome.

In the ascending groups of the fossiliferous slates of the North of England, as given in the General Section of my fourth Letter, (Wood-cut, No. 2), I do not wish to make any change: but (as the result of two short excursions I made with my friend John Ruthven, in 1851 and 1852) I can give a somewhat greater extension to the two groups (Coniston limestone and Coniston flagstone) which form the base of that section.

Secondly, in the comparative nomenclature and classification of the natural groups of Cumberland and Westmorland, I am compelled, by the progress of discovery, and with a view also to their co-ordination with the natural groups of North and South Wales, to make some not unimportant changes. These changes are made necessary, by the facts observed by myself in North and South Wales during the summer of 1846—by the publication of the great map of the Government Survey—and by the results of two short geological excursions into the Devonian and Silurian countries, made by Professor M'Coy and myself in the summers of 1851 and 1852.

Lastly, the list of Igneous Dykes (Appendix A. of the previous edition) will, I hope, be considerably increased, by a distinct enumeration of the dykes discovered by my friend John Ruthven since the spring of 1846: and the list of Fossils (Appendix B. of your last edition) will also, I trust, be augmented and improved by Professor M'Coy, in a selection of species from his great Work on the Palæozoic Fossils of the Cambridge Museum.

All these corrections and additions (addressed to the geologist rather than to the general reader) I will endeavour to describe, in the order above indicated; and, as far as possible, in a condensed and synoptical form.

I am, Dear Sir, very faithfully yours,

Cambridge, June 23, 1853. A. SEDGWICK.

I. *Extension of the three Groups which form the base of the General Section.* (Woodcut, No. 2, *supra*. p. 221).

1. These three groups are well exposed on the north side of the valley of Dent, in a section (noticed in p. 230), that commences in a deep denudation called Helm's Gill: but I was never able, before last summer, to trace the Coniston limestone and calcareous slates through the south side of the valley. My friend John Ruthven undertook the task of examining all the water-courses on the southern *strike* of the calcareous slates, and at length found some good Coniston fossils near the top of the pass leading from Dent to Kirkby Lonsdale. We subsequently discovered in the descending brows a little further north, and on the same line of *strike*, some small patches of rock with *Trinucleus Caractaci*, and other well-known Coniston fossils. Here, therefore, was the exact evidence I had been seeking for. The Coniston limestone and calcareous slates are in their right places on the south side of the valley of Dent; and over them the Coniston flags and the Coniston grits are grandly exhibited—the grits rising into the great precipices called Colm Scar.

2. In the summer of 1851, I obtained a proof of what indeed I had believed before, but on less perfect evidence: viz., that the calcareous beds immediately below the slate quarries of Ingleton beck and Thornton beck, are the equivalents of the Coniston limestone.

3. Lastly, I had a satisfactory proof, during 1851, that the upper beds of the slate rocks which appear at the base of the great scar limestone, just above Horton in Ribblesdale, are the cha-

racteristic equivalents of the Coniston limestone and calcareous slates—and that the great quarries of Moughton fell (known under the name of the "Horton flagstone"), are on the exact parallel of the Coniston flagstone.

All the facts above noticed were laid in detail before the Geological Society, and have been published in a recent volume of its Quarterly Journal. The existence of three fundamental groups (viz. Coniston limestone, Coniston flagstone, and Coniston grit) on so many parts of a line, drawn from the south end of Ravenstonedale to Horton in Ribblesdale, is an important fact in the physical history of the country on the outskirts of the Cumbrian mountains: and it is exactly analogous to what we may often remark among the old Palæozoic rocks of North and South Wales. For there, after following an ascending section almost to the outskirts of a mountain chain, we may in many places observe the beds thrown into a trough-shape, and the lower groups rising to the day, but with a reversed dip, exactly where the newest groups might, hypothetically, have been looked for. A want of attention to this fact has led to many mistakes: such, for example, as the long-continued erroneous classification of the Llandielo flags; which, in the Silurian sections, were placed over, instead of under, the upper Cambrian groups of South Wales.

II. *Classification of the Groups, &c.*

My first attempt at a classification of the 'stratified groups of the Cumbrian Mountains" was published in the spring of 1832, before I had made any great progress in disentangling the groups of North and South Wales. The "Cumbrian groups" (as I then called them) were determined on the evidence of sections. The whole series was separated into three natural divisions; and the highest division was separated into five groups. (*Proceedings of the Geol. Soc. of London*, Vol. i. p. 400, &c). Of these five groups, the lowest represented the Coniston limestone and flagstone, as given in my General Section (No. 2, *supra* p. 221), and the highest (then called "coarse greywacke and greywacke slate") represented what is now regarded as the Upper Ludlow Series, between Kendal and Kirkby Lonsdale.

Though the classification was entirely physical, the fossils did not pass unnoticed. It was then well known that the fossils of the Coniston limestone formed a different group from those of Kirkby Moor; and so far the result was satisfactory. But, while the gap between the (Upper Ludlow) rocks of Kirkby Moor and the carboniferous limestone was unexplained in any natural section, and while the Plymouth and other Devonian fossils were referred to groups of great, but unknown, antiquity, the Westmorland fossils could be turned to no account whatsoever in determining the place of the above-named five groups in a general section of British Palæozoic rocks. I mention these facts to illustrate two points. (1.) That subdivisions on good physical evidence, will, with slight modifications, often stand good after they have been submitted to the further test of fossil evidence.—(2.) That, by itself, fossil evidence is of no value in determining the place of a group in an undefined series. It can only be brought to bear *after* some good physical section has been well made out.

In the summer of 1832 I almost completed my sections through North Wales: and I then made one or two rapid and unsatisfactory traverses through the old rocks of South Wales, in the hope of making out the southern range of the Bala limestone, and of connecting, if possible, my own work with that of Sir R. I. Murchison, among certain groups of rock—afterwards called Llandeilo flag.

The general results indicated by the sections of North Wales may be thus stated:—A vast undulating series (extending from the Menai Straits to the Berwyns, and then, by great downcast faults, having its upper groups repeated on the east side of the Berwyns) might be conveniently separated into two great divisions: one (Lower Cambrian) including all the groups below the Bala limestone—the other (Upper Cambrian), commencing with the Bala limestone, and including a great series of fossiliferous slates, &c., above it. Over all this series was a great, and, in some places, discordant deposit, based on grits, sandstones, and conglomerates; and ending with the great overlying masses of the Denbigh flagstone, &c.

In this general view I had no difficulty in comparing the vast deposits of the Welsh and Cumbrian mountains. The lower Cambrian groups represented the great central Cumbrian group of green slate and porphyry.—The Bala limestone and fossiliferous slates were the equivalents of the Coniston limestone and fossiliferous slates.—The grits and conglomerates (afterwards called Caradoc sandstone) at the base of the Denbigh flags were excellent representatives of the Coniston grits. And, therefore, the great group of Denbigh flags, &c., must represent the groups of Westmorland, which range between the Coniston grits and the old red conglomerates on the banks of the Lune.

But why was not this attempt at classification published? Because the Silurian sections (which had been repeatedly discussed before the Geological Society) were thought to be directly opposed to it. I was absolutely certain that many of the undulating groups of South Wales were but the prolongation of the groups over the Bala limestone; and, therefore, so soon as the nomenclature was fixed, I called them *Upper Cambrian*—a name that may be conveniently retained. But in all the Silurian sections submitted to the Geological Society, the Llandeilo flag was placed *above* these Upper Cambrian groups: and such was the place afterwards given to it by the Author of the "Silurian System," in the map and sections of his great Work published in 1839. If these sections were true to nature, mine must have been either imperfect or false to nature; for none of my sections gave me any indication of a Llandeilo group *above* the Upper Cambrian rocks as before defined. Here was a great difficulty in the way of an accurate comparative classification of the upper groups of Westmorland and North Wales.

Nor was this difficulty removed by a visit (I made along with the Author of the Silurian System) to the best Silurian types in 1834. He still persisted in placing his Llandeilo group above my Upper Cambrian groups; and I accepted this interpretation of the sections with implicit faith, on his sole authority: for there were but very few parts of his base line which I had ever seen; and I had not critically examined any part of it. He afterwards identified the beautiful series of undulating rocks, on the east side of the Berwyns, with his Llandeilo and Caradoc groups; and he was constrained to do so by the supposed evidence of his lower sections near Welsh Pool and in Siluria. But at the same time, constrained by the irresistible evidence of the sections of the Berwyns, he admitted that the Bala limestone (spite of its group of fossils) *was* the base of a great series of rocks, which I called Upper Cambrian and which he placed below his Llandeilo group.

I accepted these conclusions, though they threw the upper groups of my own sections into inextricable confusion; and I fabricated an hypothesis to explain the conflicting phenomena. It is virtually applied in a sentence of the note affixed to my 3rd Letter (*supra* p. 201), where it is stated—"that the fossils of the Coniston and Bala limestones are very nearly the same: but the rocks are not on that account to be brought into a close comparison; because the fossiliferous rocks of North Wales (with a lower Silurian type) are of an enormous thickness, and contain bands of organic remains, some of which are far above, and some far below, the limestone of Bala." Combining this statement (which has proved true) with the hypothesis of some complicated derangements of the groups, I thought it *possible* that the groups on the east side of the Berwyns *might* be Silurian, while the Bala limestone group was Cambrian.

In 1834 I had not time to bring this hypothesis to any test; but, on revisiting North Wales in 1842 and 1843, I found that any hypothesis of the kind alluded to, was perfectly uncalled for —that my Upper Cambrian sections were right—that the Lower

Silurian sections were wrong—and that the Bala limestone and the Llandeilo flagstone must be exactly, or very nearly, on one parallel. Finally, on revisiting the sections of South Wales in 1846, I found that the Llandeilo groups were arranged in the Silurian sections, several thousand feet out of their true place in the great Cambrian series: being, in point of fact, *inferior*, instead of *superior*, to the Upper Cambrian System, as above defined. These results proved the necessity of readjusting the base line of the "Silurian System." But, instead of readjusting his base line, the Author of the System thought fit, in 1843, to expunge it altogether, and to colour all the older rocks of Wales as Silurian!

In my first published attempt, made in 1838, to compare the older deposit of the Cumbrian mountains with those of North Wales, I made the Coniston limestone contemporaneous with that of Bala, and, therefore, *Upper Cambrian*. But deceived, as I was at that time, by a belief in the integrity of the Lower Silurian Sections, and by the consequent hypothesis alluded to above, I could follow the groups no farther on distinct fossil evidence; and I only stated, in general terms, that the Westmorland rocks above the Coniston group were Silurian. I might, however, even then, have added—that the highest group was exactly, or very nearly, on the parallel of the Upper Ludlow of Siluria.

When my three Letters to Mr. Wordsworth were published (in 1842), I had not revisited any part of the great groups of North Wales, after my last attempt at a classification of them. The Letters were, therefore, written while I had still some confidence in the accuracy of the lower sections of the "Silurian System;" and while I thought it *possible*—that (on the nomenclature that had been agreed upon) the groups on the east side of the Berwyns might be lower Silurian, and the Bala bed Upper Cambrian. Hence there was, at that time, an unavoidable want of good evidence as to the exact place of the fundamental groups in the "upper division of the slate rocks." The Coniston limestone was, however, no longer supposed to represent the Bala limestone, but the Caradoc group; and with it were arranged the Ireleth slates. Of the same "division," the "upper groups" were, on far better evidence, compared with Ludlow rocks (*supra* p. 198).

My fourth Letter* appeared (in 1846) after I had made (in 1842 and 1843) two laborious tours in North Wales. My lists of fossils were greatly improved; and the groups of Howgill Fells and Middleton Fells had been at length (in 1845) arranged in their right places. The fossils of the Coniston flags seemed to be characteristic of a Wenlock group: and, as the flags passed

* Letter I. of the Supplementary Letters, in the present Edition.

by almost insensible gradations into the calcareous slates of the Coniston limestone, it seemed to follow, almost of necessity, that the Coniston limestone must continue on the parallel of the Caradoc sandstone, rather than on that of any lower group—whether Silurian or Cambrian. On this scheme was constructed the General Section (Wood-cut, No. 2), where the whole Westmorland section, from the Coniston limestone upwards, is put in co-ordination with the successive Silurian groups from the Caradoc sandstone to the Upper Ludlow rocks inclusive (*supra* p. 232).*

There was, however, one great blemish in this comparison: for the Coniston grits had no parallel among the upper groups of Siluria; while, in Westmorland, these grits had very marked features, and played such an important part, that I desrcibed them (in a Memoir read before the Geological Society in 1848—Quarterly Journal, p. 217,) as the commencement of a great physical change in the nature and colour of the deposits: and I added, in a note (p. 219), "that if our classification had been based on the Westmorland sections, we should have regarded the Coniston grits as the *commencement* of the Upper Silurian series." Moreover, the Coniston limestone was, *mineralogically*, a bad representative of the Caradoc sandstone; and was, in a part of its range, so interlaced with the green slates and porphyries of the lower system, as hardly to be separable from them. These facts threw considerable doubts upon my interpretation of the lower groups of the general section in my fourth Letter (Wood-cut, No. 2), and gradually inclined me to my first opinion—that the Coniston limestone was on the same parallel with that of Bala.

If possible to clear away these doubtful points, I re-examined, in the summer of 1851, the lower Palæozoic rocks whick break out in several well-known places between Ravenstonedale and Upper Ribblesdale; and the result was stated in a paper read before the Geological Society of London in the autumn of the same year, and since then published in the 8th volume of their Journal. To that paper I must refer for details: but I may state, that Mr. Salter, who made out my former lists, and Professor M'Coy, who has published an elaborote description of all my collection of Palæozoic fossils, now agree in opinion that the fossils of the Coniston flags do not belong to the Wenlock shale; but rather to an upper part of the Bala group. Physical and fossil evidence were thus in harmony. (1.) The Coniston limestone, calcareous slate, and flagstone (Nos. 1 and 2 of the General Section—Woodcut, No. 2), were true "Cambrian rocks," as I had first

* In 1846 I did not put the Llandeilo flag in the same group with the Caradoc.

called them; and were, therefore, to be classed along with the upper beds of the green slates and porphyries of the great central group of the Lake Mountains; and *not* with the Coniston grits and the other overlying groups. (2ndly.) The Coniston grits were apparently on the parallel of the Caradoc sandstone; and were physically the base of an upper system.

On this scheme physical and palæontological evidence seemed to be in good accordance; and the successive groups, in North Wales, above the Bala limestone, seemed to tally very nearly with the successive groups of Westmorland above the Coniston limestone. Still there remained one unexplained difficulty. Among the fossils collected by myself from the so-called Caradoc sandstone of Cambria and Siluria were two distinct groups of fossils—one of which belonged to the Wenlock—the other to the Bala type. Hence arose the following questions. Did these two fossil groups alternate? Were they so blended in the sections as to be inseparable? On the other hand, if the two groups of fossils were characteristic of two separable stages of the so-called Caradoc sandstone, to which stage must we refer the Coniston grits of the Westmorland section? If possible to clear up these questions, I made a short excursion (during the summer of 1852) to the Silurian country with Professor M'Coy; and the results of this excursion were given in a paper (read Nov. 3. 1852), which will, I hope, appear in the next number of the Quarterly Journal of the Geological Society of London.

Partly from our joint observations of the sections, and from Professor M'Coy's very careful examination of the fossils, partly also from the very important facts previously published by Professor Phillips in his elaborate Memoir (*Geological Survey*, Vol. ii. Part 1.), we arrived at this conclusion: viz. that under the name "Caradoc sandstone," two *distinct*, and often *unconformable*, groups of rocks had been confounded—the upper group containing true Wenlock—the lower true Bala fossils. To this *upper group* we gave the name of *May Hill Sandstone;* and we regard it as the true base of the Silurian System. It includes the sandstone of May Hill; the fundamental sandstone of Woolhope; nearly all the shelly sandstones on the eastern flank of the Malverns; many large tracts, with the gamboge Caradoc colour, on the great map of the Government Survey; and (as I believe from my own survey) all the sandstones and conglomerates which range, at the base of the Silurian flags of Denbighshire, from Corwen to Conway.

The May Hill sandstone is a good mineralogical representative of the Coniston grits, and the two groups of rocks occupy the same position in a general Palæozoic section. Hence I cannot but regard them as true geological equivalents. To put this conclusion beyond all doubt, requires, perhaps, the addition of a few more fossils, carefully collected from the Coniston grits. To collect fossils from these hard and sterile grits is a toilsome and ungrateful task; but I look for its performance to my friend John Ruthven, who, during a former year, went, at my request, successfully through a still harder task, and found fossils, where no one had before seen them, in the old and sterile Skiddaw slates.

The preceding historical sketch is not, I trust, out of place: for it exhibits the severe and conscientious manner in which I have endeavoured to arrive at the truth. I may still affirm, in the opening words of my fourth Letter, that "the whole series of slate rocks which extend from the centre of Skiddaw Forest to the banks of the Lune, near Kirkby Lonsdale, may be separated (as was first shown by Mr. Otley) into three great subdivisions, viz.;—

"I. Skiddaw Slate.

"II. Green Roofing Slate and Porphyry.

"III. Dark-coloured Slate and Flagstone, alternating with bands, and sometimes with thick beds of silicious gritstone."—(*Supra* p. 220).

The *Skiddaw Slate* is a deposit of vast extent; and may (at least provisionally) be compared with the old slates of the Longmynd, in Shropshire. I can find no other place for it without breaking up the co-ordination of all the older deposits of Wales and Cumberland.

The *Second Subivision*, including as it does the highest and most rugged mountains of the Lake District, must now also include the Coniston limestone and the Coniston flagstone. This classification is justified by the fossils; and puts the deposits of the Welsh and Cumbrian mountains in a better physical coordination than was effected by any previous arrangement.

The *Third Subdivision* commences with the Coniston grits (or May Hill sandstone), and includes all the remaining groups up to the old red sandstone. It therefore includes all the Upper Silurian rocks and a part of the Lower, as given in the sections of the "Silurian System;" and it includes every rock which physically and palæontologically can, with any propriety of language, be described as *a System* distinct from the Cambrian. It therefore correctly represents the "Silurian System."

If both the Skiddaw slate and the great green slate and porphyry group (as above defined) be called *Cambrian*, and all the beds, from the Coniston grits to the base of the old red sandstone, be called *Silurian*, we shall find (on an examination of Professor M'Coy's very elaborate and careful lists, derived from localities for which I can confidently vouch) that the Cambrian rocks contain seventy-two well ascertained species, and the Silurian rocks ninety-two species, and that *five species only* are common to the two; which gives us very little more than three per cent. of common species out of a total of 165. A similar per centage is, I believe, found in the magnificent Cambrian and Silurian divisions of the old rocks of North America; and the great physical and palæontological break among the older rocks of North America takes place (as I am informed by Professor H. Rogers, and collect from the great works of Hall) on the same parallel with the great break in the British series, viz.: among certain sandstones and conglomerates which initiate a new series of deposits with a new series of animal forms.

A similar per centage of common species among the Cambrian and Silurian rocks of Bohemia, may be logically derived from the fossil lists and sections of M. Barrande; and, I believe, from all other good fossil lists derived from rocks of the same age, and of which the relations are exhibited in unequivocal sections. Are, then, the fossil lists of the Cambrian and Silurian Systems of England to form an exception to all other corresponding lists, in the unusually great per centage of their common species? I do not see any good evidence for this conclusion.

While rocks, with perfect Wenlock lists of fossils, were confounded (under a common name of Caradoc) with other rocks having a characteristic list of Bala species, no wonder that all palæontological distinctions should have been abolished between the rocks of Cambria and Siluria. But when the sections are re-adjusted, and the May Hill sandstones are struck off from the sandstones of Horderley and Caer Caradoc (the latter being, of course, taken as the type of true Caradoc sandstone), I believe that the per centage of common species between true Cambrian and true Silurian rocks will be very much reduced, so as no longer to be anomalous. Physically and palæontologically the Cambrian System is more perfectly separated from the Silurian, than is the Silurian System from the Devonian, or the Devonian from the Carboniferous. That I may be understood, I here use the word System as it has of late years been current: and not in the sense in which it appears below in the Tabular view.

NOMENCLATURE OF THE OLDER PALÆOZOIC GROUPS, &c.

All good geological classification is, in the first instance, based on good unequivocal sections: and if we mean to continue our present geographical nomenclature, there can be no doubt but the names of our older Palæozoic groups must be drawn from the best Cambrian and Silurian sections. For these sections not only give us the best types of comparison, but the names derived from them have been the first that were published, and, in part accepted, by English geologists. I, therefore, think it expedient, even in this sketch, to give the following "Tabular View of the British Palæozoic System;" copied (with one single change, made necessary by the field observations of last autumn) from the advertisement to the second Fasciculus of the Palæozoic Fossils in the Cambridge Museum (July, 1852). This Tabular View gives an *ascending series*, derived, so far as regards its lower division, from the best known Cambrian and Silurian sections.

After the Granitic rocks, which appear but sparingly in Wales, we have Metamorphic and Hypozoic rocks. The Metamorphic, of great but somewhat doubtful age, and of great thickness.—The Hypozoic rocks (Longmynd, &c.) of very great thickness, and also of doubtful age; but probably to be linked to the lowest Cambrian groups, and to be placed on the general parallel of the Skiddaw Slates. If so, they will cease to be Hypozoic; and may then be considered as the lowest known base of the Cambrian Series.

Primary or Palæozoic System of Britain—in three divisions: viz. I. Lower. II. Middle. III. Upper.

I. *Lower Division, representing the Cambrian and Silurian Series in ascending groups.*

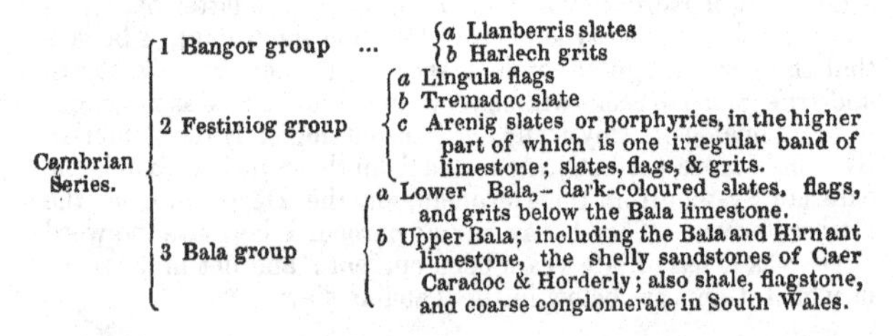

Cambrian Series.	1 Bangor group ...	*a* Llanberris slates
		b Harlech grits
	2 Festiniog group	*a* Lingula flags
		b Tremadoc slate
		c Arenig slates or porphyries, in the higher part of which is one irregular band of limestone; slates, flags, & grits.
	3 Bala group	*a* Lower Bala,—dark-coloured slates, flags, and grits below the Bala limestone.
		b Upper Bala; including the Bala and Hirnant limestone, the shelly sandstones of Caer Caradoc & Horderly; also shale, flagstone, and coarse conglomerate in South Wales.

The *Bangor group* (*a* & *b*) is of great thickness; and the Harlech grits are so well defined, in several parts of Carnarvonshire and Merionethshire, as to make good base lines on which to construct the sections. This group is without ascertained fossils: and may, perhaps, be put in co-ordination with a part of the Skiddaw slate.—The lower part of the *Festiniog group* (*a* & *b*) has no ascertained counterpart in Cumberland. The Porphyries of Arenig, Cader Idris, &c., are the perfect counterpart of the green slates and porphyries of the Lake District, and are of enormous thickness.—The *Lower Bala group* has no distinct representative in the Westmorland sections. The Upper Bala, as above stated, represents the Coniston limestone and Coniston flagstone. In North and South Wales, however, the *Upper Bala group* is expanded to the thickness of several thousand feet.—I do not think that the whole Cambrian series, represented by the above three groups, is much less than 30,000 feet in thickness.

Lower Division continued.

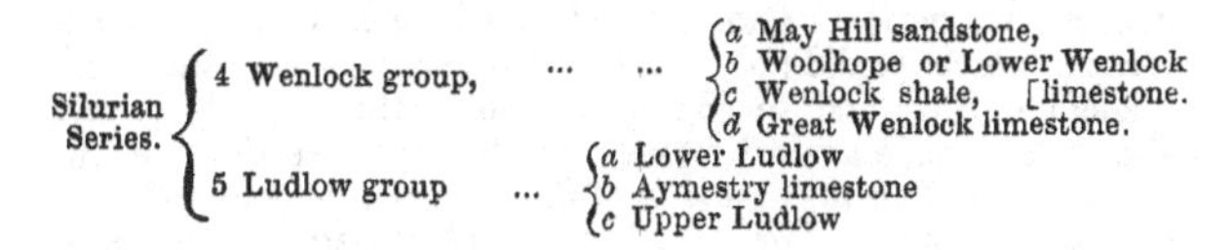

Silurian Series.	4 Wenlock group, ...	*a* May Hill sandstone, *b* Woolhope or Lower Wenlock limestone. *c* Wenlock shale, *d* Great Wenlock limestone.
	5 Ludlow group ...	*a* Lower Ludlow *b* Aymestry limestone *c* Upper Ludlow

The May Hill sandstone is computed, by Phillips, at about 1000 feet. The Coniston grits (its equivalent?) are considerably thicker. The thickness of the other sub-groups may be obtained from published authorities.

II. *Middle Division.*

Devonian Series.	6 Plymouth group	*a* Plymouth limestone and red grit; Liskeard slate, &c. *b* Dartmouth slate,
	7 Hereford or Caithness group	*a* Dipterous flags, &c, *b* Hereford sandstone, marl, & cornstone
	8 Petherwin group	*a* Marwood sandstone, &c. *b* Petherwin slate and Clymenia limestone

III. *Upper Division.*

Carboniferous Series.	Divisible into three or four groups.
Permian Series.	Divisible into three groups.

In the above Tabular View, the only change, in the arrangement of the groups, which differs considerably from that which was first adopted by Sir R. I. Murchison and myself, in 1834, is the necessary removal of the Llandeilo flag and sandstones of Caer Caradoc and Horderley, &c., into the Cambrian series. But, by far the greatest part of the rocks formerly named Caradoc, and lately coloured as Caradoc in the great map of the Government Survey, are here regarded as the true base of the Silurian series, and are grouped under the name—May Hill sandstone. The reader who is acquainted with any geological map that was constructed on the principles of nomenclature given in the map of the "Silurian System" (1839), will see, that the proposed arrangement makes but a small change in the distribution of colours on the maps of North and South Wales, and of the adjacent English counties. The original scheme of geographical nomenclature remains in its integrity—the rocks of Cambria are called Cambrian; the rocks of Siluria are called Silurian;—and every group of the "Silurian System," the true relations of which had been discovered by the author of the "System," retains its name and place in the general sections.

When, in 1834, I visited North and South Wales, along with Sir R. I. Murchison, I then, for the first time, examined his best Silurian groups, which I adopted without reserve: and as for the demarcation between Cambria and Siluria he had a perfect *carte blanche* which we did not examine or discuss together, and which I adopted from end to end on his sole authority. I did this, although it made against myself, and threw my sections at the north end, and on the east side, of the Berwyns, into almost inextricable confusion. For I had, in 1832, constructed a series of lower Cambrian sections up to the Bala limestone: and a section from the Bala limestone to the north end of the Berwyns, where the upper Cambrian rocks were overlaid by a coarse sandstone (afterwards called Caradoc): and these sections *appeared* to represent the *whole Cambrian series*, up to the base of an *upper series*, afterwards called Silurian. These very sections were exhibited and discussed in 1838 (before the publication of the "Silurian System"), and are sufficiently described in the "Proceedings of the Geological Society" (Vol. ii. p. 679).* But the upper section was rejected, because it gave no place for the Llandeilo group; which, in the Silurian sections of my fellow-labourer, was placed *over* the upper Cambrian groups. My section was, however, right in principle and right in detail: but I had no power to parry the objections taken to it (by all who discussed the point, except Professor Phillips), so long as I believed, on the authority of the Author of the "Silurian System," that the Llandeilo flags were *superior* to my upper Cambrian groups.

As for the *Bala limestone*, there *could* be no doubt about its place in the Welsh series; and on the clear evidence of the sections, and in spite of its fossils, it was placed far within the west, or Cambrian, side of the line of demarcation adopted in the "Silurian System." In short, my friend decided, on the spot, that the Bala limestone *could not be brought within the limits of his "System,"* and he never communicated to me any subsequent change in this opinion. The previous statement admits not of contradiction; for it is literally true. Yet in the "Proceedings

* See also "Proceedings of the Geological Society" Vol. iii. p. 548—550, (1841).

of the Geological Society" (Nov. 1842, vol. iv. p. 10), it is gravely affirmed, as if it were an allowed truth, "that Professor Sedgwick stated the Upper Cambrian System (Bala limestone, &c.) to lie below the Silurian System of Mr. Murchison—*a view adopted by Mr. Murchison, upon the authority of Professor Sedgwick*"!

Every word I wrote respecting the Cambrian groups, from the summer of 1834, to the summers of 1842 and 1843, when I revisited North Wales, was naturally affected by the erroneous position that had been given by my friend, to his Lower Silurian groups. But, during the summers of 1842 and 1843, I found that on no reasonable hypothesis could the fossiliferous rocks (of Meifod, &c.) on the east side of the Berwyns be separated from the Upper Bala group—and that my original sections through the northern end of the Berwyns, and all my lower Welsh sections, were perfectly right in principle, and generally right in detail. It followed, that the groups of Meifod, the older groups near Welsh Pool, and, probably, all the Llandeilo groups, must be at *the base*, and not, as in the "Silurian System," at the top of the groups I had called Upper Cambrian; and I then adopted an hypothetical opinion, that nearly all the conglomerates, slates, &c. of South Wales, which I had called Upper Cambrian, would prove to be Upper Silurian.

While labouring under this uncertainty (not at all arising from any original mistake of mine), I thought that the Upper Cambrian and Lower Silurian groups were inseparably entangled; and to prevent the unnecessary repetition of ill-defined names, I adopted the name of *Protozoic group*, to define at once all the fossiliferous groups of Wales that were known either as Cambrian or Lower Silurian. ("*Proceedings*," vol. iv. p. 223, June, 1843).

Of a Paper in the first volume of the Quarterly Journal of the Geological Society (Nov. 29, 1843), I can hardly claim the authorship. It is, in fact, an abridgement of two of my previous papers, by a former President; and though I applied to him, again and again, he (for reasons best known to himself) did *not allow me*, during its passage through the press, to see *a single proof-sheet of the reduced Memoir!* The consequence was, that in the letter-press and sections there are several errors; and that the small accompanying map is very imperfect and inaccurate. But of this I do not so much complain, as of the names given, upon the map, to the superficial colours: for there the word *Protozoic* is made synonymous with *Lower Silurian;* in *direct violation* of the whole spirit of my papers.* I do not ac-

* That the reader may understand the unwarranted liberty taken (in ignorance and with no sinister intention) with my paper, let him turn to the Pro-

cuse the President of treachery. I verily believe that in thus *Silurianizing* my paper he thought he was doing me a favour: but I trust that a like unwarranted liberty will never again be taken with an author's works, so long as the Geological Society shall exist.

Lastly, in 1846, I once more made traverses through the undulating series of South Wales; and, after an interval of twelve years, I revisited the Llandeilo groups. The final result was a full conviction that my Upper Cambrian groups were correctly placed and named in my original sections; and that the true relations of the Landeilo group, both to the bed above it and below it, had from the first, been misunderstood and misrepresented in the "Silurian System." There was, therefore, no shadow of reason for greatly changing my original nomenclature; and the Llandeilo group, the place of which had been mistaken, must at length rest in its true place, near the base of the upper Bala groups of the preceding Tabular view.

I then regarded the Caradoc group (which in many places appeared to be unconformable to the Cambrian rocks) as the true base of the System of Siluria. On this scheme there was a good physical, but not a good palæontological base to that System. But, as before stated, that last difficulty has now almost vanished, by the establishment of a May Hill group, distinct from the Caradoc.

It is impossible for any one to be a judge of the controversy between Sir R. I. Murchison and myself, without some knowledge of the facts above stated. So far from intruding on my friend's province, I almost superstitiously respected every group that he had professed to make his own; and I endeavoured to lend support to his fundamental sections, which were wrong, by vainly torturing, out of their natural meaning, my own upper sections, which were right. On the contrary, when he found that his assumed Silurian System was likely to be shaken (not merely because it had no base, but because its supposed base was dislocated and useless), he stole a march upon me, and spread his lower Silurian colour over all the old Cambrian groups.* When a controversy arose afterwards between us, he justified the change

ceedings of the Geol. Soc. (vol. iv. p. 223), where I define the term *Protozoic*, and state that it includes "all the older fossiliferous slates of North and South Wales, Coniston limestone, lower part of the 'Silurian System' of Mr. Murchison," &c. The term *protozoic* was adopted for the *express purpose* of avoiding any collision between the Lower Silurian groups and the Upper Cambrian. Whether there were any real overlap between the true Silurian and Upper Cambrian groups was then considered doubtful; but it was certain that the position of some of the so-called Lower Silurian groups had been entirely mistaken.

* The first intimation I had of this astonishing extension of the Silurian colours was from Mr. Knipe, some time, if I mistake not, in 1845.

by an appeal to a section the base of which was incorrigibly erroneous; and how can an enduring nomenclature be drawn from a section, that is not true to Nature? There ought never to have been any controversy between us.

The Author of the Silurian System has always appeared to argue as if all his groups and names must remain untouched, and all other groups and names be discarded; although his Lower Silurian names are neither geographically appropriate, nor drawn from sections which were understood when his System was published. On the contrary, the names of the Cambrian rocks, which he does his best to discard, are geographically appropriate, and are drawn from true sections. When not misled by a paternal partiality to his own nomenclature no man living can see his way more clearly. Thus (Feb. 1842) he wrote as follows: "So long as British geologists are appealed to as the men whose works in the field have established a classification founded on the sequence of the strata and the embedded contents, so long may we be sure that their insular names, &c., will be honoured with a preference by foreign geologists." Most true! so long, and no longer. So far as my friend has established his groups, and their nomenclature, both on true sections and good groups of fossils, his names are accepted by myself and all English geologists. But I never did accept his names, after he had transgressed his own principles, and endeavoured to extend his names indefinitely beyond the limits of his true sections; and thereby virtually endeavoured to throw the imputation of error on his fellow-labourer, while the error was all his own.

The flagstones of Builth and Llandeilo, with *Asaphus Buchii*, were well known to collectors, and I had obtained specimens from them a dozen years before we either of us entered Wales. By the noble list of fossils from these localities, my friend did good service to palæontology: but by mistaking the geological position of the Llandeilo group in the general Cambrian Series, he threw a very formidable stumbling-block in the way of further progress. On this point I can speak from unfortunate experience. And even had he laid down this group in its right place in the general section of Wales, it would I, think, have been but a sorry reason for claiming, as his own, full 20,000 feet of strata of which he knew almost nothing from personal observation; and which I had already named, after no small personal labour, compatably with the evidence of the sections, and with true geographical consistency; and, I might almost say, in conformity with his express wishes.

In 1834, the question between us was one mainly of sections, and not one exclusively of fossils: and spite of its fossils, which (as proved in his great work, published in 1839) were all of known Lower Silurian species, he rejected the Bala group from his

System. If he afterwards saw reason to change that opinion, the change ought to have been communicated to myself. I objected to the word System again and again, as applied to the Silurian groups; because the System had no well-defined base—either physical or palæontological. And when my friend began to discover that his System was in danger, as a System; he had no right to make good his over-ambitious language at my expense; and because his own boundary line was insecure, to fortify it by invading a province to which he had no rightful claim on the ground of previous conquest, or priority of discovery.

But my friend now took up a new position, and contended that fossils were all in all, and that the Cambrian series must disappear from our nomenclature because its fossils were Silurian. While attempting to reach this conclusion, he seemed to forget that I had, on very good grounds, objected to his first adoption of the word *System*, as applied to his collective Silurian groups: and that to surrender the Llandeilo group to the great Cambrian series (after it was found that the position of the group had been mistaken), was far more natural and rational, (and I might add, far more graceful,) than a vain attempt to merge all Wales in Siluria. Had there been any well-established section, worked out both on physical and fossil evidence, to which both the Cambrian and Silurian groups could be made co-ordinate, this argument, from fossils only might have been plausibly maintained. But there was no such ascertained typical section. The whole series of groups was new;—neither the Cambrian nor the Silurian groups were perfectly made out; and while such was the condition of our knowledge, the final nomenclature must remain in, at least, partial abeyance; and the bare fossil argument, for purposes of true scientific nomenclature, was perfectly worthless.

But there is what seems to me a strange (though I doubt not unintentional) misrepresentation in some of my friend's published statements. His readers must have fancied that I knew nothing of the fossils of Cambria; and that I was, in 1842 and 1843, seeking, through Mr. Salter's help for a *Systema Naturæ*, among the older Welsh groups, quite distinct from that of Siluria. As the human muscles will sometimes produce rotatory movements without any conscious exercise of the will, so the human mind will sometimes reach a "foregone conclusion" without any remembrance of previous facts, or any exercise of the inductive faculty. Thus, my friend, in his Anniversary Address, ("Proceedings," Vol. IV. p. 73. Feb. 1843.) wrote as follows: "We were, both aware that the Bala limestone fossils agreed with the Lower Silurian; but *depending upon Professor Sedgwick's conviction* that there were other, and inferior masses, also fossiliferous, we both *clung to the hope* that such strata, when thoroughly explored would offer a sufficiency of new forms to characterize an inferior System."—I smiled when I read this strange passage; but I did not think it worth while formally to contradict it.—In omission and commission it is a virtual misstatement of the facts. The Author does not inform the reader, that he had himself consented, in 1834, to put the Bala limestone in my Upper Cambrian groups. —Because "it had a sufficiency of new forms" to mark a new System? By no means: but because it was the base of a great physical group which he himself had excluded from his own System in South Wales; and *over which* he had erroneously (as was afterwards made out by other observers) placed his Llandeilo group. Nor does he tell the reader that I had from the first strenuously opposed the adoption of the word *System*, when applied to the Silurian groups; because they had no defined base either physical or palæontological. The sentence quoted proves, to demonstration, that my original objection (and I may add the repeated objections of Professor Phillips) to the word System had been right;—that the Silurian nomenclature was still in abeyance;—and that it must be considerably modified in order to bring it into any conformity with a true geographical nomenclature, and with the palæontological evidence of more complete sections.

There were excellent Silurian groups; but there never was a "Silurian System" such as the author published. Not merely because it had, at the time of its publication, no good base; but also (as now appears certain) because he had himself overstepped its true base—the May Hill sandstone. What was originally wrong could not be made right (against all claims of priority and all geographical propriety) by an unheard-of process of "downwards development," and by an incorporation of the vast and well-ascertained groups of Cambria, into his comparatively insignificant and misinterpreted lower groups of Siluria.

Again, when the Author states—"that we both clung to the hope, that the Cambrian groups would offer a sufficiency of new forms to characterize an inferior System," I can only reply—that *the hope to which he clung* was not derived from anything I had ever said or written; and that I had not, in 1842 and 1843, the shadow of a hope that any new system of animal life—any group, of "new forms marking an inferior System"—would be found among the lower Cambrian groups. I had constantly expressed and frequently published, a *directly contrary opinion;* and it was *one of the grounds* of my strong objection to the word System, as applied to the Silurian groups. Thus, in a *published* Syllabus of my Lectures drawn up in 1836, I naturally gave a synopsis of the "Silurian System," so far as it was at that time known—enumerating its groups and some of their fossils. And to preserve a symmetry of language (though at the time disliking the word

System, as then applied) I divided the whole Cambrian Series into two Systems—Upper and Lower Cambrian—which were merely assumed as subdivisions convenient for description; but were not adopted with any implied reference whatsoever to peculiar groups of fossils: one term simply implying all Cambrian groups above, the other all Cambrian groups below, the Bala limestone. In 1836, the great mass of my collections was inaccessible to myself: but sometime after 1834 I had submitted a small reserved collection, of common and characteristic species from the Bala limestone, to Mr. Sowerby; and the result is recorded in my Syllabus (p. 51). "The limestone contains *Bellerophon bilobatus Producta sericea, and nine species of Orthis*—all of which are common to the lower Silurian System." There was not so much as one shell in the small collections (which, with the corals, made up about twenty species) that could be pronounced by Mr. Sowerby distinct from shells in the Silurian M.S. lists.—Again, at the end of the synopsis of the *Upper Cambrian rocks,* in the same Syllabus, were these words—"*Many shells of the same species with those of the Lower Silurian rocks.*"

Again, in the still lower Cambrian groups, between the Bala limestone and Arenig, were multitudes of fossils (of which I had no reserved collection), which seemed to carry on the same general type. Of the fossils from the Carnarvon trough that runs through the top of Snowdon I had a small reserved collection which contained, as I believed, only Bala species. To prevent mistake, it was submitted to Mr. Sowerby, in 1841—the year *before* I revisited North Wales—and the result ("Proceedings of the Geol. Soc." vol. iii. p. 549) only confirmed my previous belief. Every ascertained species had its match in the Bala group.

Again, in 1838, ("Proceedings." vol. ii. p. 679) writing of the Upper Cambrian groups, I stated that "many of its fossils are identical in species with those of the lower division of the Silurian System; nor have any true distinctive zoological characters of the group been well ascertained." Again, in 1841, ("Proceedings" Vol. III. p. 548) the same statement is repeated almost in the same words. From all this evidence it is beyond dispute, that in the passage, above quoted, in the President's Address, my views were strangely misunderstood, and entirely misrepresented.

After the unsanctioned abridgement of my papers, and the *unwarranted change* of my nomenclature, before alluded to, I am compelled to refer again to a paper by myself, of which I did make the abridgement ("Proceedings," vol. iv. June 1843). I there (p. 221) state my opinion respecting the *Cambrian and Lower Silurian,* fossiliferous rocks, in the following words, "The fossiliferous series above described is called the "great protozoic group of North Wales." There is no good fossil evidence for its separation into distinct formations; and its inferior beds, although far below the Caradoc sandstone, contain comparatively few species undescribed in the work of Mr. Murchison. It is therefore neither Cambrian nor Silurian in the limited sense in which the words were first used: but it represents both Systems (that is, all fossiliferous Lower Silurian and Cambrian groups), inseparable as they are in nature from one another." The paper concludes with the following classification of British Palæozoic rocks. They are considered zoologically as one system, separable into four *primary divisions*, as follows. (1). Carboniferous and Permian.—(2) Devonian—(3) Silurian—including only the Upper Silurian rocks of Mr. Murchison.—(4) Protozoic, as above defined. The reader will perceive, at once, how nearly this scheme agrees with that of the above Tabular View.

Misled by the speech of the President, from which I have already quoted ("Proceedings" vol. iv. p. 73); and probably still more misled, by the first paper in the Quarterly Journal of the Geological Society, and especially by the names written by the President on the accompanying map (without any sanction from myself, and in direct antagonism with my own views,) the Director of the Government Geological Survey believed that I had virtually abandoned my nomenclature after the year 1843. This may partly account for the nomenclature of the older palæozoic groups of Wales since adopted by the gentlemen of the Government Survey. On their scheme a great and ancient group of rocks (Longmynd, &c.) is called exclusively Cambrian, though its best type is in England.—Twenty or thirty thousand feet of strata magnificently seen in Wales, hardly seen in Siluria, and, when seen, misunderstood in the Silurian System—are to be called Silurian—And all Siluria and Cambria is to make one Silurian System! Their work, I doubt not, is of consummate perfection in most of its details; but in repudiating my proposed nomenclature they seem to have mistaken their own position; and to have forgotten that all my sections are right in principle; and that there is not one single important group in North Wales which I had not described and put in its right place before them. I believe that the sketch of a general classification given by myself in 1843 ("Proceedings of the Geol. Soc." vol. iv. p. 223) however imperfect, is far truer and more philosophical than their own; which is at once geographically incongruous and historically unjust. I believe that thier classification also involves a great Palæontological mistake. We can hardly over-value the importance of a good nomenclature. The gentlemen of the Survey, misled by the adopted name Caradoc, have confounded together things that should be separated, and have thereby missed the true base of an upper (or Silurian) series: and following the same lead, they have linked together, in a false union, two vast groups of rocks

(Silurian and Cambrian) which nature elaborated under very different physical conditions, and during very different periods of animal life. I can still adopt the words published by myself "ten years since ("Proceedings" Vol. iv. p. 224), and affirm with truth, " that the two great divisions (viz. Cambrian and Silurian) differ in structure, interchange hardly any fossil species, and, through large districts are unconformable. Hence they belong to two *systems,* and not to one, if the word *system* be used in a definite sense, and be applied to the successive *divisions* of the palæozoic rocks, such as the Devonian."

But all the successive palæozoic divisions are so intimately united, and, however they may differ in fossil species, have so much of a common type, that they seem to form a distinct *Systema Naturæ.* Hence, I have given a unity to the Palæozoic System, and think that the word *system* may be retained conveniently in that extended sense. I have used the word Series, because it formerly passed current to describe large groups of rocks (such as Cretaceous series, Oolitic series, &c.) *before* the word *system* was a word of common use among British Geologists in the description of similar groups. Among foreign geologists the word *terrain* often stands in the place of *system,* as used by Sir R. I. Murchison; and their word *système* has generally been applied, to subordinate rather than to principal groups. I have never been ambitious to give new names to groups of strata; and I am certain that premature nomenclature does much mischief, and hinders men from looking nature honestly in the face. All I am truly anxious about is, that I may assist in urging on the course of truth, without injustice to any fellow labourer; and that what I have done may help to take away some mistakes in fact, classification, and nomenclature, that have grievously hindered her progress.

CONCLUSION.—More than thirty years are gone since I first endeavoured to disentangle the structure of the Cumbrian Mountains. Should I again have the happiness of visiting them, it must be under the consciousness of enfeebled powers and diminished strength; no longer allowing me to grapple with tasks that once called forth the exertion of willing and hopeful labour. Thoughts such as these would temper my enjoyment in a country, rich, though it be, in natural beauty, and associated with the bright remembrances of youthful life:—and other thoughts would also arise fitted, for a time, to sink the heart with sorrow.

It was near the summit of Helvellyn that I first met DALTON —a truth-loving man of a rare simplicity of manners; who, with humble instruments and very humble means, ministered, without flinching, in the service of high philosophy, and by the strength of his own genius won for himself a name greatly honoured among all the civilized nations of the earth.—It was, also, during my geological rambles in Cumberland that I first became acquainted with SOUTHEY, that I sometimes shared in the simple intellectual pleasures of his household, and profited by his boundless stores of knowledge. He was, to himself, a very hard task-master: but on rare occasions (as I learnt by happy experience) he could relax the labours of his study, and plan some joyful excursion among his neighbouring mountains.—The friends I have named are gone; "but their memorial shall never be forgotten."

Most of all, during another visit to the Lakes, should I have to mourn the loss of W. WORDSWORTH; for he was so far a man of leisure as to make every natural object around him subservient to the habitual workings of his own mind; and he was ready for any good occasion that carried him among his well-loved mountains. Hence it was that he joined me in many a lusty excursion, and delighted me (amidst the dry and sometimes almost sterile details of my own study) with the out-pourings of his manly sense, and with the beauteous and healthy images which were ever starting up within his mind, during his communion with nature, and were embodied, at the moment, in his own majestic and glowing language.

It would be idle for me to criticise his works. In any labour, however humble, I think it a high honour to have my own name associated with his; for he was, in very truth, a great and good man, and a lasting benefactor to that delightful country in which he spent nearly all the latter years of his long life. That he worked out, and not without many a well-fought battle, a great revolution in the taste and poetical sentiments of Englishmen, admits of no doubt: and in his published volumes he has laid open a goodly storehouse of happiness for those who will take him for their guide, and will listen to his song.

But while the imagination is set free from one error, it will sometimes fall into another; and there have been men who, after a long poetical communion with the outer world, have learned, at length, to be idolaters of Nature to the verge of Pantheism: so as to hold cheap the social duties of life; and duties higher still, and aspirations far more noble than ever sprang within the soul of man from any imaginative intercourse with the outer glories of Nature. Wordsworth, from the time I first knew him, was not merely pure from habit and self-control; but he was pure from the influence of a principle that soared above any motives which he drew from his communion with Nature. He was a man of firm religious convictions; "and many a time, when it was my great happiness to roam with him over his native mountains, have I heard him pour out his thanks, that, while he had been permitted to slake his innermost thirst at Nature's spring, he had been led to think of the God of Nature, and not to forget His redeeming love." Let, then, no wanderer among the Lakes, "while he honours the great Poet who is gone, forget the Poet's faith: or dare to draw from his noble lessons the materials of an idolatrous or pantheistic dream, and then to tell us that he is of the school of Wordsworth."* With this quotation, and not without a feeling of sorrow, I now bring my letters to a close.

* "Discourse on the Studies of the University of Cambridge" (1850). Supplement to the Appendix, p. 318.

APPENDIX (A).

ON THE PORPHYRY DYKES WHICH TRAVERSE THE SLATE ROCKS.

A short notice is token, in my third letter,* of the dykes of igneous rock in the several divisions of the slate series. The following notice (with the exception of the dykes of Black Coomb and the valley of the Duddon) is confined to the dykes in the fossiliferous division of the slate series described in my fourth letter. All minute details are omitted, and the following statement professes to be little more than a bare enumeration of the places where the phenomeno may be studied.

I. *Black Coomb and the valley of the Duddon.*—This dislocated region (bounded to the north by the Bootle granite, is traversed by dykes at the following places: and beyond doubt, several more dykes will be discovered, and many must remain concealed under the turf bogs.

1. One or two granular felspar dykes break out through the black "Skiddaw slate," in the brows on the south side of Black Coomb. Not far from one of them is a mineral vein.

2. A great dyke, 60 or 70 feet wide, is seen on the S.E. side of Black Coomb at Whitchamp Mill, above Beckside. It ranges nearly with the beds of "Skiddaw slate," and mineralizes them only to the depth of a few inches.

3. A similar dyke appears farther up the mountain stream. The two last mentioned are said to traverse the neighbouring hills of "green slate and porphyry," but I have not followed them on their strike.

4. A fine dyke of "quartziferous porphyry" is seen in one of the extreme branches of Green Gill, on the N. side of Black Coomb.

5. Another (about a mile east of Fenwick farm), near the highest point of the junction of the "Skiddaw slate" with the "green slate and porphyry."

(The following are from the second division of the slate series, "green slate and porphyry.")

6. A red felspar dyke, near Smalthwaite farm, and S. of the mountain road from Duddon Bridge to Bootle.—7. & 8. Two similar dykes, crossing the same road, about half a mile north of Fenwick.—9. A dyke (perhaps a continuation of No. 6.) of bright red felspar rock, crossing the Duddon between Pen and Sallow; it may be followed nearly parallel to the strike of the slates, for about two miles.—10. Another, in the bed of the river, about half a mile below Ulpha Church.—11. Another, near the saw-mill, south-west of the church; and may, perhaps, be continued to No. 12,—which breaks out near Whinfield Ground.—13. Lastly, at Wallabarrow (below Seathwaite), where a magnificent red felspar dyke rises from the river, through the crag (in a vertical mass which seems to thin away in its ascent). It may thence be followed for more than two miles, nearly on the bearing of the strike, to the bed of the rivulet under a farm called Brow Side, where it disappears.

The prevailing structure of these dykes is that of a red granular felspar rock, with subordinate grains of quartz: but the quartz sometimes predominates, and becomes highly crystalline, and the rock passes into a quartziferous porphyry. At Whitchamp Mill, the great dyke (No. 2) is extremely pyritous, and some of the specimens containing small flakes of mica, become granitic. In

* Page 211, and Note, page 213.

addition to their beautiful mineral structure, these dykes derive an interest from their association, as above stated, with some of the greatest dislocations that have affected the Cumbrian mountains.

II. *Dykes in the fossiliferous slates between Windermere and Duddon Sands.*—1. A fine red felspar dyke, which may be traced in a direction nearly from Field Head, near Hawkshead, to the Coniston road under Hawkshead Hill. At its S. end it appears to branch out into two distinct dykes.

2. Another crosses the road a little above Hawkshead Hill, and may be followed along a water-course about half a mile to the S. W. It appears to strike nearly with the *Coniston flags*, through which it breaks.

3. A third, of similar structure, and twenty feet wide, may be traced for about 100 yards through the *Ireleth slates*, just skirting the *Coniston grits*. The spot is nearly defined by the intersection of two lines—one drawn N.E. from Esthwaite Old Hall—the other N. by W. from Sawrey Chapel. Numerous bowlders of red syenitic rock, especially on the brows which rise on the S.W. side of Esthwaite Water, may, perhaps, indicate the passage of one or more dykes towards Grizedale.

4. A fourth breaks out at the top of the brow, N.E. of Grizedale Hall, and crosses the valley near the village (where it is buried under the superficial drift). It then breaks out by the side of the mountain road leading to Coniston Water foot, and may be followed, almost continuously, to the road on the side of the lake, nearly opposite to the island, and it probably crosses to the other side of the water, near Brown Hall. This dyke strikes for several miles with the beds; but does not dip with them. In one or two places it underlies, at a great angle to the N.W., while the slate beds dip S.E. Perhaps the direction of the fissures through which the dykes were poured, may have been more or less modified by previously existing planes of slaty cleavage.

5—8. The country between Brown Hall and Tottlebank Fell is traversed by four or five lines of dykes, which appear to intersect one another, and may only be ramifications of one central trunk. All of them have nearly the same structure: their chief constituent being red felspar; and some specimens show obscure green spots, derived from green earth, or, perhaps, from hornblende. In the low hills immediately W. of Brown Hall, are three lines of great felspar rock. The central mass, which is 30 or 40 feet wide, strikes in a direction about S.S.W. across the brook that runs down from Beacon Tarn. Again, to the W. of this brook, masses of similar rock break out upon, at least, two lines—one ranges on the N. side of Tottlebank Fell—the other parallel to, and on the south side of the thin band of *Ireleth limestone* (4*b*).

9. A dyke of the same prevailing structure (for a knowledge of which I was first indebted to the Rev. J. Bigland, of Finsthwaite) appears to be continuous for six or seven miles nearly along the strike of the "*Ireleth slates.*" Its range is defined by the following places, where it is well exposed to view. The hill top at the S. side of the Graythwaite Hall plantations, High Crag Head, above Thwaite Head, Force Beck, the meeting-house above Rusland Chapel, below Ickenthwaite, Abbot's Park, and Hill Park farm. It probably runs down the hill into the Bridge Field estate, where there are numerous bowlders of the rock, but no open quarry. The further range of this great dyke is cut off by the valley of Crake.

10. A fine dyke about a mile above the chapel, and at the S. end of the village of Rusland, is seen on the road side. It mineralizes the beds in contact with it, and probably ranges in a direction nearly parallel to the former dykes, but it has not been traced far, and its direction is therefore doubtful.

11. Lastly, a dyke of decomposing felspar rock, or porphyry, breaks out on the side of the high road between Cartmel and Backbarrow, a little S. of the road turning off to Bigland Hall. Its range is not defined in the quarry, but it is said to reappear to the S.W. near Stribers.

All the dykes last enumerated (from No. 3 to No. 11) traverse the zone of "*Ireleth slate.*" They have a common prevailing structure; and, in a few places, decompose into great flakes which are parallel to the sides of the dykes, and give them the deceptive appearance of red slates irregularly interpolated among the other rocks. This structure (sometimes seen in true granite near its junction with a slate rock), was probably produced during the slow cooling of the mass, whereby the sides of the dykes passed gradually into a solid state sooner than their central portions.

III. *Dykes in the fossiliferous slates of Westmorland, &c.**—1. A dyke on the N. side of Wet Sleddale, nearly opposite the great boss of Shap granite, cuts through the green slate series, and therefore does not properly belong to this last.

2. A beautiful dyke, rising through the steep hills on both sides of the valley which descends to High Borrow Bridge from the N.W. It is about 17 feet wide, is nearly vertical, crosses the valley near High House, and strikes (like the beds through which it runs) nearly N.E.; and parallel to its sides it shows the kind of slaty structure before noticed. None of the remaining dykes show this structure in any perfection.

3 & 4. A strong dyke of quartziferous porphyry on the E. side of the road over Shap Fells, not far from the Demmings. It strikes towards the N.W., and perhaps runs into another dyke (No. 4), which crosses the ancient road ascending the hills from Hause Foot.

5. Another dyke, 160 yards long, and 36 yards wide, is seen near the top of Knot, nearly E. of Borrow Bridge.

6. A very fine syenitic mass breaks out on both sides of the fell-gate above Brestye, on the road from High Borrow Bridge to Bretherdale.

7. A similar rock, but ill-exposed and decomposing, is seen in a water-course descending from Whinfell Common to Borrowdale, above Low Borrowdale Farm: and also in the following places, viz.—8. In a water-course (on the south side of Whinfell Common, and not marked in any map) which descends to Forest foot, above half a mile N. of Grayrigg fell-gate.—9. In the side of a brook which descends from the highest point of the turnpike to the Lune, beyond Dillicar Knot.—10. In a brook a little E. of Lummer Head farm.—11. The road side, a little E. of Lambrigg fell-gate.—12. In a water-course about a quarter of a mile to the south of 11.—13. Again, at Foster How, at the source of a brook which runs through New Hutton Park.—14. At Dubside, three-quarters of a mile W. of 13.—15. The mountains of Howgill and Ravenstonedale are much covered up, but a few rolled fragments of porphyry (probably derived from dykes) have been found in the water-courses. A decomposing dyke does however occur in Bretherdale, nearly opposite Yarlside.

16. All the above dykes derive their light, and, generally, red colour from felspar. But in the cuttings for the Lancaster and Carlisle Railway, about two miles S. of Low Borrow Bridge, under the highest part of Dilicar Knot, is exposed a dyke of dark-coloured amygdaloidal greenstone (a very rare appearance in this country), protruded among some ferruginous and shattered strata.

17. Holmescale Fell—may be seen running through a quarry, and appearing in two or three places near the Old Hutton Parsonage House.

18. In the side of a well at Gillow-Style, Firbank; and, again, a little to the N.E. in a water-course, where the dyke crosses the brook six times.

19. Dromer Stile, near Birthwaite,—well exposed in the Railway cutting, crossing the line six times, and mineralizing the rocks through which it passes.

20. Staveley Head, between High Gillbank and Low Gillbank—Barley Bridge—New Hall Estate, about a quarter of a mile S.W. of which the dyke disappears in Crook.

21. At Cocks Close, E. side of Potter Fell—may then be traced S.W. to near Godmond Hall—again in the river below Cowan Head—at Boston—at Rather Heath, crossing the Ambleside road near the third milestone—then striking S.W. appears near Low Birks, Crosthwaite—at Cartmel Fell, near the school—and near Fox Field.

22. Sturdy Crag, near Aikrigg End, Kendal.

All the dykes up to No. 10 inclusive, and Nos. 15, 16, 19, 20, & 21, occur in the Second and Fourth Divisions of the General Section.

No. 11 is in the 5th, and Nos. 12, 13, 14, 17, 18, & 22 are in 5*b*. of the General Section.

IV. *Dykes of Yorkshire, near the line of the great Craven fault.*—1–3. Commencing on the confines of Ravenstonedale, about 200 yards below Rother Bridge, and descending the river about the same distance, three dykes are seen in its bed. The highest appears to send off a lateral branch about 20 feet wide: and two re-

* Four or five of the dykes enumerated under this and the following head were first discovered by Mr. John Ruthven, Kendal.

markable bosses of red felspar rock are so immediately associated with these dykes, that they all apparently belong to one system of eruption.

4. A dyke of similar structure to the preceding, breaks out in three places among some dislocated strata which are partly covered by the conglomerates of the old red sandstone, in a brook which comes down to Hall Beck farm, on the S. side of Garsdale foot.

5—11.—On entering the common above Helm's Gill, in Dent, we meet with seven dykes crossing the brook within the distance of about 200 yards. The most southern of the group (No. 5) is about 40 feet wide, and within it a mass of the calcareous shale is entangled; it also sends strings or small branches into the slates. Nos. 6, 7, 8, are only from two to five feet in thickness; No. 9 is about 30 feet thick; No. 10 of small thickness. The preceding six dykes are so nearly parallel to the true beds that they might be taken for regular strata without close examination. But they are followed by No. 11, a beautiful red vertical dyke which ranges nearly at right angles to the mean strike of the other dykes, and of the beds.

It is probable that there are many more dykes along the same line; for no rock is here visible except along the channel of one small water-course which springs from beneath extensive turf bogs.

12. Another breaks out on the S. side of the valley of Dent, in a brook which descends from Holm Fell, and falls into the river Dee below Rash Mill.

13. The preceding dyke may, perhaps, be continued over the ridge of Holme Fell to a point a little S. of Brown Knot, where another dyke of similar structure comes to the surface through a thick covering of heath.

14. A dyke at Colm Scar.

15. Lastly, one or two dykes break through the calcareous slates (Coniston limestone?) about 200 yards below the Ingleton slate quarries.

All the dykes enumerated under this fourth head of the Appendix have nearly the same structure. They are essentially composed of felspar and black mica; and the latter mineral in broad flakes is sometimes so abundant as to form the largest constituent of the rock. They are generally granular, and rarely pass into a compact rock exhibiting a base of felstone with interspersed small spangles of mica. Granular quartz never appears to predominate among these dykes, as it does among some of those enumerated under the preceding heads. Their external colour is yellowish red, but in their interior they sometimes exhibit a dark bluish tint. Considering their peculiarity of structure, we may suppose that they were produced under one set of conditions; and from the proximity of most of them to the great Craven fault, we might perhaps conjecture that they were protruded during the action of those disturbing forces which formed the 'Craven fault.' If so, they are of a date posterior to the carboniferous groups. On the other hand, not one of them appears to cross the line of the Craven fault, or to pierce any rocks of the carboniferous age. Hence their epoch may be considered doubtful; especially when we bear in mind that the faults which have broken into fagments the carboniferous zone on the south-western skirts of the Lake mountains, and greatly deranged the highest slate groups, are not marked by the presence of igneous dykes. Whatever may be decided respecting the dykes here considered, it is, I think, almost certain that nearly all the dykes, described under the three preceding heads of this Appendix, took their place among the aqueous rocks before the period of the old red sandstone.

Specimens of the preceding dykes, labelled with their localities, have been collected and deposited in the Kendal Museum, by Mr. John Ruthven. Specimens are also deposited in the Woodwardian Museum, Cambridge, by Professor Sedgwick.

NOTE. While these sheets were passing through the press, Mr. J. Ruthven has discovered another Dyke, west of Wray, in Docker Parks, striking nearly N. and S.

LIST OF FOSSILS

DERIVED FROM LOCALITIES (IN CUMBERLAND, WESTMORLAND, AND PARTS OF LANCASHIRE AND YORKSHIRE) ALLUDED TO IN THE PREVIOUS LETTERS.

Cambrian Fossils, from the Skiddaw slate to the Coniston flag inclusive.

GRAPTOLITES, latus (M'Coy)
„ sagittarius (His. sp.)
„ Ludensis (Murch.)
PALÆOPORA, interstincta (Wahl. sp.)
„ Var. subtubulata (M'Coy)
„ megastoma (M'Coy)
„ petalliformis (Lonsd. sp.)
„ tubulata (Lonsd. sp.)
FAVOSITES, crassa (M'Coy)
NEBULIPORA explanata (M'Coy)
„ papillata (M'Coy)
STENOPORA fibrosa (Gold. sp.)
„ Var. *a* Lycopodites (Say)
HABYSITES catenulatus (Linn. sp.)
SARCINULA organum (Linn. sp.)
PETRAIA æquisulcata (M'Coy)
BERENICEA heterogyra (M'Coy)
PTILODICTYA explanata (M'Coy)
RETEPORA Hisingeri (M'Coy)
CARYOCYSTITES Davisii (M'Coy)
TENTACULITES annulatus (Schlof.)
BEYRICHIA strangulata (Salt)
LICHAS subpropinqua (M'Coy)
CERAURUS clavifrons (Dal. sp.)
ZETHUS atractopyge (M'Coy)
ODONTOCHILE obtusi-candata (Salt.)
PORTLOCKIA apiculata (Salt.)
CHASMOPS Odini (Eichw. sp.)
CALYMENE brevicapitata (Portk.)
„ sub-diademata (M'Coy)
HOMALONOTUS bisulcatus (Salt.)
ISOTELUS Powisi (Murch.)
ILLÆNUS Rosenbergi (Eichw.)
SPIRIFERA biforata (Schlot.)
„ Var. *b.* dentata (Pand. sp.)
„ Var. *d.* fissicostata (M'Coy)
„ insularis (Eichw. sp.)
SPIRIFERA per-crassa (M'Coy)
PENTAMERUS lens (Sow. sp.)
ORTHIS Actoniæ (Sow.)
„ calligramma (Dal.)
„ crispa (M'Coy)
„ expansa (Sow.)
„ flabellulum (Sow)
„ parva (Pander)
„ plicata (Sow. sp.)
„ porcata (M'Coy)
„ protensa (Sow.)
„ Vespertilio (Sow.)
LEPTÆNA deltoidea (Conrad)
„ Var. *b.* undata (M'Coy)
„ minima (Sow.)
„ sericea (Sow)
„ transversalis (Dal.)
STROPHOMENA antiqua (Sow. sp.)
„ grandis (Sow. sp.)
„ pecten (Linn. sp.)
„ spiriferoides (M'Coy)
LEPTAGONIA depressa (Dal.)
LINGULA Davisi (M'Coy)
„ ovata (M'Coy)
PTERINEA termistriata (M'Coy)
CARDIOLA interrupta (Brod.)
ORTHOCERAS filosum (Sow.)
„ laqueatum (Hall)
„ vagans (Salt.)
„ subundulatum (Portk.)
„ tenuicinctum (Portk.)
CYCLOCERAS annulatum (Sow. sp)
„ Ibex (Sow.)
„ subannulatum (Munst.sp.)
LITUITES cornuarietis (Sow.)
Total 72.

Silurian Fossils, from Coniston grits to Upper Ludlow rock inclusive.

NEBULIPORA papillata (M'Coy)
STENOPORA fibrosa (Gold. sp.)
„ Do. Var. *b.* regularis (M'Coy)
HALYSITES catenulatus (Linn. sp.
CYATHAXONIA Siluriensis (M'Coy)
SPONGARIUM æquistriatum (M'Coy)
„ interlineatum (M'Coy)
„ interruptum (M'Coy)
ACTINOCRINUS pulcher (Salt.)
TAXOCRINUS orbigni (M'Coy)
ICTHYOCRINUS pyriformis (Phill. sp.)
URASTER primævus (Forb.)
„ Ruthveni (Forb.)
„ hirudo (Forb.)
PROTASTER Sedgwickii (Forb.)
TETRAGONIS Danbyi (M'Coy)
CORNULITES Serpularis (Schlot.)
TENTACULITES tenuis (Sow.)
SERPULITES dispar (Salt.)
TRACHYDERMA squamosa (Phill.)
BEYRICHIA Klodeni (M'Coy)

CERATIOCARIS elliptica (M'Coy)
„ inornata (M'Coy)
„ solenoides (M'Coy)
ODONTOCHILE candata (Broug. sp.)
„ Var. minor
CALYMENE tuberculosa (Salt.)
HOMALONOTUS Knighti (Konig.)
FORBESIA latifrons (M'Coy)
EURYPTERUS cephalaspis (Salt.)
SIPHONOTRETA Anglica (Morris)
DISCINA rugata (Sow. sp.)
„ striata (Sow.)
SPIRIFERA sub-spuria (d'Orb.)
SPIRIGERINA reticularis (Linn. sp.)
HEMITHYRIS navicula (Sow. sp.)
„ nucula (Sow. sp.)
ORTHIS lunata (Sow.)
STROPHOMENA filosa (Sow. sp.)
CHONETES lata (V. Buch. sp.)
LINGULA cornea (Sow.)
AVICULA Danbyi (M Coy)
PTERINEA Boydi (Conrad sp.)
„ demissa (Conrad sp.)
„ lineata Gold.)
„ pleuroptera (Conrad)
„ retroflexa (Wahl. sp.)
„ var. naviformis
„ subfalcata (Conrad sp.)
„ tenuistriata (M'Coy)
CARDIOLA interrupta (Brod.)
MODIOLOPSIS complanata (Sow. sp.)
„ solenoides (Sow. sp.)
ANODONTOPSIS angustifrons (M'Coy)
„ bulla (M'Coy)
„ securiformis (M'Coy)
ORTHONOTUS semisulcatus (Sow. sp.)
SANGUINOLITES anguliferus (M'Coy)
„ decipiens (M'Coy)
LEPTODOMUS amygdalinus (Sow. sp.)
LEPTODOMUS globulosus (M'Coy)
„ truncatus (M'Coy)
„ undatus (Sow. sp.)
GRAMMYSIA cingulata (His sp.)
„ Var. *b*. triangulata (Salt)
„ Var *g*. obliqua (M'Coy)
„ extrasulcata (Salt. sp.)
„ rotundata (Salt.)
ARCA Edmondiiformis (M'Coy)
„ primitiva (Phill.)
CUCULLELLA coaretata (Phill. sp.)
„ ovata (Sow. sp.)
NUCULA Anglica (d'Orb.)
TELLINITES affinis (M'Coy)
CONULARIA cancellata (Sandb)
„ subtilis (Salt)
PLEUROTOMARIA crenulata (M'Coy)
MURCHISONIA torquata (M'Coy)
NATICOPSIS glaucinoides (Sow. sp.)
HOLOPELLA cancellata (Sow. sp)
„ gregaria (Sow. sp)
„ intermedia (M'Coy)
LITORINA corallii (Sow. sp.)
„ octavia (d'Orb. sp.)
BELLEROPHON expansus (Sow.)
ORTHOCERAS angulatum (Wahl.)
„ baculiforme (Salt.)
„ bullatum (Sow.)
„ dimidiatum (Sow.)
„ imbricatum (Wahl.)
„ laqueatum (Hall)
„ sub-undulatum (Portk.)
„ tenuicinctum (Portk.)
CYCLOCERAS Ibex (Sow.)
„ subannulatum (Munst. sp.)
„ tenuiannulatum (M'Coy)
„ tracheale (Sow. sp)
HORTOLUS Ibex (Sow. sp.)
Total 98.

The previous List, drawn up by Professor M'Coy, has the following abbreviations, which it may be well to explain :—

His. Hisinger.—Murch. Murchison.—Wahl. Wahlemberg.—Lonsd. Lonsdale. —Linn. Linnæus.—Schlot. Schlotheim.—Salt. Salter.—Dal. Dalman.—Eichw. Eichwald.—Portk. Portlock.—Pand. Pander.—Sow. Sowerby.—Brad. Braderip. —Munst. Munster.—Gold Goldfuss.—Phill. Phillips.—Forb. Forbes.—Broug Brougmiart.—d'Orb. d'Orbigny —Sandb. Sandberger.

The letters *sp.* of course, means *species:* and where a name is written with sp. after it, the symbols mean—that the author quoted gave the name of the species, but with a different generic name.

BOTANICAL NOTICES.

Alchemilla *alpina*.—Above Buckbarrow Well, Long Sleddale, and near the smmit of Helvellyn, and Lake Mountains.
Allium *arenarium*.—By the river side near Helsington, and Mint Bridge, near Kendal.
——— *oleraceum*.—Borders of Derwentwater, and near Kirkby Lonsdale Bridge.
——— *Schœnoprasum*.—Rusmittle, Lyth, near Kendal.
Anchusa *sempervirens*.—Near Tolson Hall gate, near Kendal, and by the road-side in the Vale of Long Sleddale.
Andromeda *polifolia*.—On Brigsteer Moss, near Kendal.
Apium *graveolens*.—On Brigsteer Moss, near Kendal.
Aquilegia *vulgaris*.—At foot of Brigsteer Scar, near Kendal.
Arbutus *Uva-Ursi*.—Descending Grassmoor to Crummock Water.
Arabis *petrœa*.—Screes, near Wastwater.
Arenaria *verna*.—Above the Lime Kilns, Kendal Fell.
Asarum *europœum*.—About Keswick.
Asperula *cynanchica*.—Abundant on Kendal Fell.
Asplenium *viride*.—On the edge of Scout Scar, near Kendal.
Aspidium *aculeatum, var. lonchitiforme*.—At Scarfoot, near Kendal.
——— *oreopteris*.—Stony places near Long Sleddale.
Astragalus *glycyphyllus*.—On rocks at Humphrey-head, near Cartmel, and Cul-garth Pike, Keswick.
Atropa *Belladonna*.—About Furness Abbey.
Bidens *tripartita*.—Near Burneside Hall, near Kendal.
Botrychium *lunare*.—In meadows near Barrowfield Wood, and Singleton Wood, near Kendal.
Brachypodium *Sylvaticum*.—Cunswick Wood, near Kendal.
Calamintha *officinalis*.—Kendal Castle.
Campanula *latifolia*.—In the hedges, about Heversham near Milnthorp and Kendal.
——— *trachelium* —In Park Head Lane, near Kendal.
——— *glomerata*.—Hardendale, near Shap.
Cardamine *amara*.—Laverock Lane, near Kendal.
Carduus *nutans*.—Near the Toll-bar, Shap.
Carex *vesicaria*.—About Cunswick Tarn, near Kendal.
Carex *rigida*.—Skiddaw and Helvellyn.
Chrysosplenium *alternifolium*.—Near the Gate at Benson Hall, near Kendal.
Cicuta *virosa*.—About Keswick.
Circæa *alpina*.—On the road-side between Ulverston and Hawkshead, and on the margins of Derwentwater.
Cladium *mariscus*.—About Cunswick Tarn, near Kendal.
Clinopodium *vulgare*.—Cunswick Wood, near Kendal.
Cnicus *heterophyllus*.—Peat Lane, Kendal, Hardendale, near Shap, and Long-sleddale
Cochlearia *officinalis, var. Grœnlandica ?*—Above Buckbarrow Well, Longsleddale.
Coronopus *Ruellii* —Beast Banks, Kendal.
Corydalis *claviculata*.—In Spital Wood, near Kendal.
Colchicum *autumnale*.—Mintsfeet, near Kendal, and Greenside, Milnthorpe.
Comarum *palustre*.—Skelsmergh Tarn, near Kendal.
Convallaria *multiflora*.—At Holker, near Cartmel, Castlehead Wood, near Keswick, and near Grange.
——— *majalis*.—Cunswick Wood, near Kendal.
——— *Polygonatum*.—Barrowfield Wood, near Kendal, but rare.
Conyza *squarrosa*.—In Levens Park, and at Scout Scar, near Kendal.
Convolvulus *Arvensis*.—Near Heversham.
Cryptogramma *crispa*.—Peat Lane, Kendal, and above Buckbarrow Well, Long-sleddale.
Cystopteris *fragilis*.—Gilling-grove, Kendal, and Kendal Fell.
Cynoglossum *officinale*.—Near Levens Church.

Drosera *rotundifolia* }
——— *longifolia* } On Fowlshaw Moss.
——— *Anglica* }
——— *longifolia.*—Near the seventh milestone on the road from Kendal to Ambleside.
Daucus *carrota.*—Abundant on Kendal Fell.
Epipactis *latifolia.*—Cunswick Wood, near Kendal.
——— *palustris.*—About Cunswick Tarn, near Kendal.
——— *ensifolia.*—Barrowfield Wood, near Kendal, but rare, and Woods at Lowther and Grange.
——— *grandiflora.*—Woods at Lowther, opposite Askham Hall.
Epilobium *alsinifolium.*—Above Buckbarrow Well, Longsleddale.
——— *angustifolium.*---By the River Side above High Borough Bridge.
Equisetum *hyemale.*—Near Old Field Wood, near Kendal, by the river side.
Euonymus *europœus.*—Near Hundow, Kendal.
Eupatoria *canabinum.*—About Cunswick Tarn, near Kendal.
Festuca *ovina, var. vivipara.*—Above Buckbarrow Well, Longsleddale.
Galium *boreale.*—Under Kirkby Lonsdale Bridge, and at Hardendale, near Shap.
——— *pusillum* —Abundant on Kendal Fell.
Galeopsis *versicolor.*—Sprint Bridge, and Burneside Hall, near Kendal.
Gentiana *Pneumonanthe.*—On Fowlshaw Moss, near Grange.
——— *Amarella.*—Above the Lime Kilns, Kendal Fell.
——— *campestris.*—Above the Lime Kilns, Kendal Fell.
Geranium *Sylvaticum.*—Coniston Water Head, and common in most of the wooded lanes near Kendal.
——— *phœum.*—Between Kirkby Lonsdale and Cowan Bridge, at Keswick, and at Pepper Hag, near Burneside.
——— *sanguineum.*—Scout Scar, near Kendal.
——— *robertianum, white var.*—In a field near Jenkin Crag, Kendal.
——— *columbinum.*—Near Fell-foot, Newby Bridge, and Canal Banks, Kendal.
——— *pyrenaicum.*—Keswick
Geum *rivale.*—Laverock Bridge, near Kendal.
Gnaphalium *dioicum* —On Kendal Fell, on high pastures in Longsleddale, and Wastdale Screes.
Grammitis *Ceterach.*—Near Fell-side, Crosthwaite, and on Kendal Fell.
Gymnadenia *conopsea.*—Rusmittle, Lyth, and Cunswick Tarn, near Kendal.
Habenaria *albida.*—On the high ground between Coniston and Hawkshead, and about Wathendlath Tarn, and at Barrowfield Wood, but rare.
——— *bifolia* —Rusmittle, Lyth, near Kendal
——— *chlorantha.*—Cunswick Tarn.
——— *viridis.*—Tenter-fell, Stricklandgate, Kendal.
Helianthemum *canum.*—On rocks at Humphrey-head, near Cartmel, and Scout Scar, near Kendal.
Helleborus *Viridis.*—In a field on the left side of Banrigg farm-house, near the eighth milestone from Kendal to Ambleside.
Hesperis *matronalis.*—Rivulets about Dale Head, Thirlmere.
Hieracium *paludosum.*—In a marsh behind Spittal Wood, Kendal, and in several moist situations.
——— *Lawsoni.*—Between Shap and Anna Well.
Hippocrepis *comosa.*—Scout Scar, near Kendal.
Hottonia *palustris* —On Brigsteer Moss, near Kendal.
Hymenophyllum *Wilsoni.*—Nook, Ambleside, and Sna Cave, Longsleddale.
Hypericum *androsœmum.*—About the Ferry, Windermere.
——— *dubium.*—Below Kirkby Lonsdale Bridge.
——— *montanum* }
——— *hirsutum* } Scout Scar, near Kendal.
——— *elodes* —Near the seventh milestone on the road from Kendal to Ambleside.
Hyoscyamus *niger.*—Near Levens Church, near Kendal.
Hypochœris *maculata.*—On rocks at Humphrey-head, near Cartmel.
Impatiens *noli-me-tangere.*—Stock Gill Force, Ambleside.
Inula *Helenium.*—Fell-side farm, Crosthwaite, near Kendal.
Juncus *triglumis.*—Fairfield, and West Side of Helvellyn.

Juncus *filiformis* —Foot of Derwentwater.
Lathræa *squamaria*.—In Levens Park, Cunswick Wood, and Laverock Bridge, near Kendal
Lepidium *Smithii*.—Near Lodore, Keswick.
Litorella *lacustris*.—About Derwent Water.
Luzula *pilosa*.—Laverock Bridge, near Kendal.
——— *spicata*.—Fairfield mountain.
Lycopus *europæus*.—Burneside Farm, near Kendal.
Lycopodium *selaginoides*.—Above Buckbarrow Well, Longsleddale.
Malva *Sylvestris*.—Near Heversham, Milnthorpe.
Melampyrum *sylvaticum*—Whitbarrow Woods.
Mecanopsis *cambrica*,—Peat Lane, Oxenholme, Sprint Bridge, near Kendal, near the Chapel, Longsleddale, and about the Ferry, Windermere.
Menyanthes *trifoliata*—Common in Tarns.
Mentha *rotundifolia* —Between Lodore and Bowderstone.
Meum *athamanticum*.—Docker Garths, The Green, and Lambrigg Fell gate near Kendal.
Monotropa *Hypopitys*.—Barrowfield Wood, near Kendal
Myrrhis *odorata*.—About Spittal, near Kendal.
Narcissus *Pseudo-Narcissus*.—Pine Crags and Ratherheath, near Kendal.
Nuphar *lutea*.—Near the seventh milestone on the road from Kendal to Bowness.
Nymphœa *alba*.—Ditto.
Ophrys *muscifera*.—Rusmittle, Lyth, and Barrowfield Wood, near Kendal.
——— *Nidus Avis*.—Cunswick Wood, near Kendal.
Ophioglossum *vulgatum*.—In meadows near Barrowfield, and in Singleton Wood, near Kendal.
Orchis *latifolia and maculata*.—About Cunswick Tarn, near Kendal.
——— *ustulata* —About Keswick.
Ornithopus *perpusillus* —On the road side on the east of Coniston Lake, and Tenterfell, Kendal.
Origanum *vulgare* —Cunswick Wood, near Kendal.
Osmunda *regalis* —By the road side under Whitbarrow
Oxyria *reniformis*.—Above Buckbarrow Well, Longsleddale, and Black Rocks of Great End.
Paris *quadrifolia*.—In Spittal Wood, near Kendal.
Parnassia *palustris*.—About Cunsick Tarn, near Kendal.
Peucedanum *Ostruthium*.—By a brook from the north end of Thirlmere.
Polypodium *vulgare, var. Cambricum*.—In Levens Park.
——— *calcareum*.—Whitbarrow and Kendal Fell.
——— *Dryopteris*.—At Scarfoot, and near the fifth milestone on the road from Kendal to Ambleside, and Singleton Woods.
——— *Phegopteris* —At Scarfoot and Stock Gill Force; and in Singleton Woods, near Kendal.
Polygonum *viviparum*.—Hardendale near Shap.
Potentilla *verna* —Whitbarrow Woods.
——— *fruticosa*.—In the Devil's Hedge-gate, Wastdale Screes.
Poterium *Sanguisorba*.—Scout Scar, Kendal, and Hardendale Nab, near Shap.
Primula *elatior* —Cunswick Wood and Spittal Wood, near Kendal.
——— *farinosa*.—About Cunswick Tarn, near Kendal.
Prunus *Padus*.—In Spittal Wood, near Kendal.
Pyrus *Aria* —On rocks at Humphrey-head, near Cartmel; and Scout Scar, near Kendal.
Pyrola *media*.—Stock Gill Force, Ambleside.
Pyrola *Secunda* —Between Great Dodd and Helvellyn.
Ranunculus *auricomus*.—Laverock Lane, near Kendal.
Rhamnus *catharticus*. } Cunswick Wood, near Kendal.
——— *Frangula*. }
Rhodiola *rosea*.—On the sides of Goatscar, Longsleddale.
Ribes *alpinum*.—Docker Brow, near Kendal.
Rosa *bractescens*.—Ambleside.
Rubus *saxatilis*.—Cunswick Wood, near Kendal.
Salix *Smithiana* }
—— *Weigliana* } On the banks of the Lune, near Kirkby Lonsdale.
—— *tenuifolia* }
—— *Croweana* }

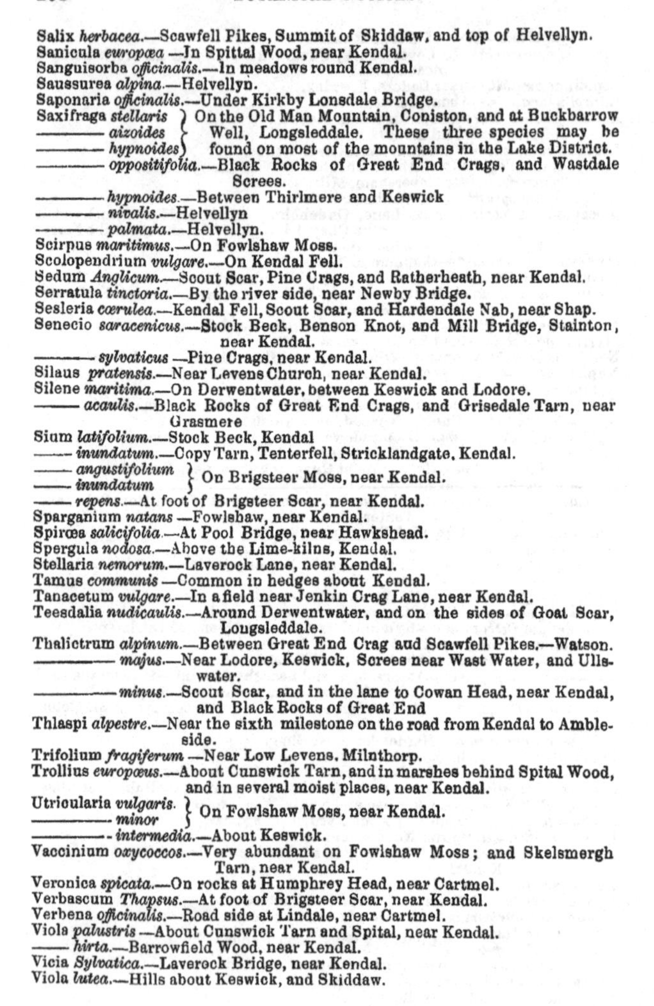

Salix *herbacea*.—Scawfell Pikes, Summit of Skiddaw, and top of Helvellyn.
Sanicula *europæa*.—In Spittal Wood, near Kendal.
Sanguisorba *officinalis*.—In meadows round Kendal.
Saussurea *alpina*.—Helvellyn.
Saponaria *officinalis*.—Under Kirkby Lonsdale Bridge.
Saxifraga *stellaris* } On the Old Man Mountain, Coniston, and at Buckbarrow
——— *aizoides* } Well, Longsleddale. These three species may be
——— *hypnoides* } found on most of the mountains in the Lake District.
——— *oppositifolia*.—Black Rocks of Great End Crags, and Wastdale Screes.
——— *hypnoides*.—Between Thirlmere and Keswick
——— *nivalis*.—Helvellyn
——— *palmata*.—Helvellyn.
Scirpus *maritimus*.—On Fowlshaw Moss.
Scolopendrium *vulgare*.—On Kendal Fell.
Sedum *Anglicum*.—Scout Scar, Pine Crags, and Ratherheath, near Kendal.
Serratula *tinctoria*.—By the river side, near Newby Bridge.
Sesleria *cœrulea*.—Kendal Fell, Scout Scar, and Hardendale Nab, near Shap.
Senecio *saracenicus*.—Stock Beck, Benson Knot, and Mill Bridge, Stainton, near Kendal.
——— *sylvaticus*.—Pine Crags, near Kendal.
Silaus *pratensis*.—Near Levens Church, near Kendal.
Silene *maritima*.—On Derwentwater, between Keswick and Lodore.
——— *acaulis*.—Black Rocks of Great End Crags, and Grisedale Tarn, near Grasmere
Sium *latifolium*.—Stock Beck, Kendal
——— *inundatum*.—Copy Tarn, Tenterfell, Stricklandgate, Kendal.
——— *angustifolium* } On Brigsteer Moss, near Kendal.
——— *inundatum* }
——— *repens*.—At foot of Brigsteer Scar, near Kendal.
Sparganium *natans*.—Fowlshaw, near Kendal.
Spirœa *salicifolia*.—At Pool Bridge, near Hawkshead.
Spergula *nodosa*.—Above the Lime-kilns, Kendal.
Stellaria *nemorum*.—Laverock Lane, near Kendal.
Tamus *communis*.—Common in hedges about Kendal.
Tanacetum *vulgare*.—In a field near Jenkin Crag Lane, near Kendal.
Teesdalia *nudicaulis*.—Around Derwentwater, and on the sides of Goat Scar, Longsleddale.
Thalictrum *alpinum*.—Between Great End Crag and Scawfell Pikes.—Watson.
——— *majus*.—Near Lodore, Keswick, Screes near Wast Water, and Ullswater.
——— *minus*.—Scout Scar, and in the lane to Cowan Head, near Kendal, and Black Rocks of Great End
Thlaspi *alpestre*.—Near the sixth milestone on the road from Kendal to Ambleside.
Trifolium *fragiferum*.—Near Low Levens, Milnthorp.
Trollius *europœus*.—About Cunswick Tarn, and in marshes behind Spital Wood, and in several moist places, near Kendal.
Utricularia *vulgaris*. } On Fowlshaw Moss, near Kendal.
——— *minor* }
——— *intermedia*.—About Keswick.
Vaccinium *oxycoccos*.—Very abundant on Fowlshaw Moss; and Skelsmergh Tarn, near Kendal.
Veronica *spicata*.—On rocks at Humphrey Head, near Cartmel.
Verbascum *Thapsus*.—At foot of Brigsteer Scar, near Kendal.
Verbena *officinalis*.—Road side at Lindale, near Cartmel.
Viola *palustris*.—About Cunswick Tarn and Spital, near Kendal.
——— *hirta*.—Barrowfield Wood, near Kendal.
Vicia *Sylvatica*.—Laverock Bridge, near Kendal.
Viola *lutea*.—Hills about Keswick, and Skiddaw.

LIST OF

LAND AND FRESH-WATER SHELLS

FOUND IN THE NEIGHBOURHOOD OF KENDAL, OR WITHIN A FEW MILES.

NERITINA *fluviatilis* (River Neritine).—In the Lune, near Kirkby Lonsdale.
BITHINIA *tentaculata* (Tentacled Bithinia).—Canal. Brigsteer Moss.
VALVATA *piscinalis* (Stream Valve Shell).—Brigsteer Moss. Large in Castle Mills race.
„ *cristata* (Crested Valve Shell)—Brigsteer Moss.
ARION ater (Black Arion).—Abundant.
LIMAX *maximus* (Spotted Slug).—In gardens and outhouses.
„ *agrestis* (Milky Slug).—In gardens.
VITRINA *pellucida* (Transparent Glass Bubble). Under stones. Not uncommon.
HELIX *aspersa* (Common Snail)—In gardens. Too common.
„ *hortensis* (Garden Snail)—In gardens and hedges.
„ *nemoralis* (Girdled Snail)—In gardens and hedges. Large on Kendal fell.
„ *arbustorum* (Shrub Snail)—Canal banks, and about the Castle.
„ *pulchella* (White Snail)—On a garden wall at Green Bank, and many other places.
„ var. *costata*—Among moist moss near Sizergh Fellside.
„ *fulva* (Top-shaped Snail)—Serpentine Walks, Kendal—Hyning wood.
„ *aculeata* (Prickly Snail)—Near Beck Mills, Low Groves, and Oxenholme.
„ *hispida* (Bristly Snail)—Kendal fell, and under stones in shady places.
„ *concinna* (Neat Snail)—On Kendal fell, under stones.
„ *rufescens* (Rufous Snail)—Abundant.
„ *caperata* (Black-tipped Snail)—Kendal fell. Morecamb Bay. Not com.
„ *ericetorum* (Heath Snail)—On Kendal fell. Common. [fell.
ZONITES *rotundatus* (Radiated Snail)—Common under stones. Large on Kendal
„ *umbilicatus* (Open Snail)—On Kendal fell. Abundant.
„ *alliarius* (Garlic Snail)—Serpentine Walks.
„ *cellarius* (Cellar Snail) Do. and many other places.
„ *nitidulus* (Dull Snail) Do.
„ *crystallinus* (Crystalline Snail)—Serpentine Walks. Hyning wood. Not common.
SUCCINEA *putris* (Common Amber Snail)—In the Canal.
BULIMUS *obscurus* (Dusky Twist Shell)—On Kendal Fell and Sizergh Fellside.
ZUA *lubrica* (Common Varnished Shell)—Under stones. Common.
AZECA *tridens* (Glossy Trident Shell)—On Kendal fell. Not common.
ACHATINA *Acicula* (Needle Agate Shell)—At Arnside. Not common.
PUPA *umbilicata* (Umbilicated Chrysalis Shell)—Kendal fell. Very common.
„ *juniperi* (Juniper Chrysalis Shell)—Kendal fell. Abundant.
„ *marginata* (Margined Chrysalis Shell)—Kendal fell.
VERTIGO *edentula* (Toothless Whorl Shell)—On Kendal fell. Not common.
„ *pygmæa* (Pigmy Whorl Shell)—On Kendal fell. Old walls. Common.
„ *substriata* (Six-toothed Whorl Shell)—Serpentine Walks. Scarce.
„ *pusilla* (Wry-necked Whorl Shell)—Hawes bridge, Serpentine Walks, and near Mint House. Not common.
„ *alpestris* (Alpine Whorl Shell)—Kendal fell.

Balæa *perversa* (Fragile Moss Shell)—On old walls, among moss. On a wall near Fowl Ing. Very abundant on old walls near Bowness.

Clausilia *bidens* (Laminated Close Shell)—Kendal fell, Helsfell wood, in a fence near Madgegill, Helsington, and Arnside Knot. Abundant near Boundary Bank.

" *nigricans* (Dark Close Shell)—On old walls and trees,—on the Castle walls. Common.

Carychium *minimum* (Minute Sedge Shell)—Hyneing wood, and Cunswick, on decayed leaves a few inches under ground.

Limnæus *auricularius* (Wide-mouthed Mud Shell)—Canal?

" *glaber* (Eight-whorled Mud Shell)—Ellerflat Tarn, near Docker Garths.

" *pereger* (Puddle Mud Shell)—Canal, Kent, and brooks. Common.

" *palustris* (Marsh Mud Shell)—In the Kent, below Water Crook, and on Brigsteer Moss.

" *truncatulus* (Ditch Mud Shell)—In ditches, and in wet places near the Limekilns.

Amphipeplea* *glutinosa*—Glutinous Membrane Shell)—In Windermere.

Ancylus *fluviatilis* (Common River Limpet)—In rivulets. Common. Large in the Canal.

Velletia *lacustris* (Oblong Lake Limpet)—On Benson Knot.

Physa *fontinalis* (Stream Bubble Shell)—Canal, Brigsteer Moss, & Windermere.

Planorbis *albus* (White Coil Shell)—Canal, Brigsteer Moss, and Milldam at Cowan Head.

" *imbricatus* (Nautilus Coil Shell)—Copy Tarn and Brigsteer Moss.

" *carinatus* (Carinated Coil Shell)—Brigsteer Moss.

" *marginatus* (Margined Coil Shell)—Do.

" *vortex* (Whorl Coil Shell) Do

" *spirorbis* (Rolled Coil Shell)— Do. and Aikrigg Tarn.

" *contortus* (Twisted Coil Shell)—Brigsteer Moss.

Cyclostoma *elegans* (Elegant Circle Shell)—At the roots of fern on Arnside Knot

Cyclas *cornea* (Horney Cycle)—In a ditch near Helm Lodge, & Brigsteer Moss.

Pisipium *pusillum* (Minute Pera)—In the Kent, and Brigsteer Moss.

Anodon *cygneus* (Swan Fresh-water Muscle)—Canal.

" var. 8 *anatina*—Brigsteer Moss.

Alasmodon *margaritiferus* (Pearly Alasmodon)—In the Mint, Kent, and Gowan.

* It occurs in Windermere Lake, but in one part only, according to Mr. Winstanley.—(Forbes and Hanley's *Brit. Mollusca.*)—It has not been ascertained in what part of the Lake this shell occurs; the specimens examined were taken from the stomach of a large Windermere trout.

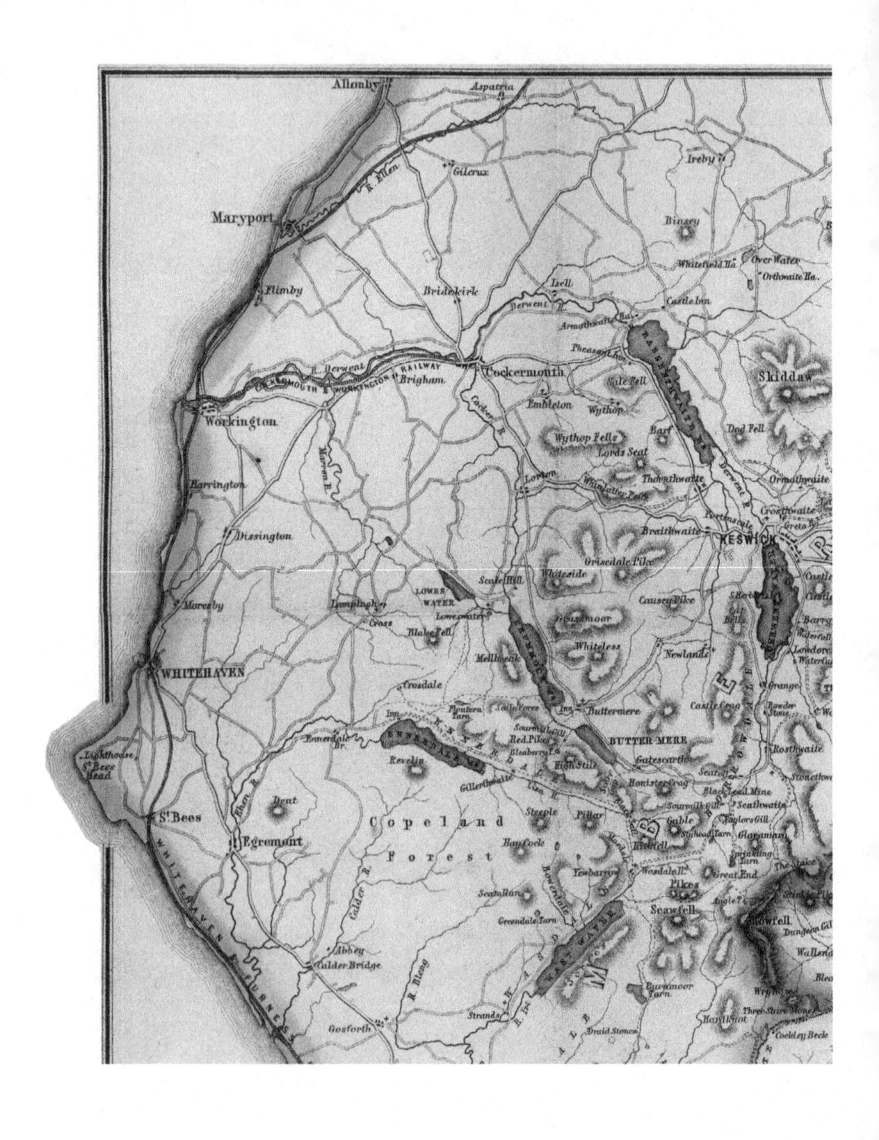

Allonby
Aspatria
Ireby
R. Ellen
Gilcrux
Maryport
Binsey
Over Water
Orthwaite Ha.
Whitefield Ha.
Flimby
Bridekirk
Isell
Derwent
Castle Inn
Pheasant Inn
R. Derwent
COCKERMOUTH & WORKINGTON RAILWAY
Brigham
Cockermouth
Skiddaw
Workington
Embleton
Wythop
Wythop Fells
Lords Seat
Barf
Dod Fell
Harrington
Lorton
Thornthwaite
Ormathwaite
Crosthwaite
Portinscale
Braithwaite
KESWICK
Dissington
Grisedale Pike
Whiteside
Scale Hill
LOWES WATER
Lowes water
Lamplugh
Cross
Blake Fell
Moresby
Causey Pike
Newlands
Whiteless
Mellbreak
WHITEHAVEN
Crosdale
Grange
Castle Crag
Bowder Stone
Buttermere
BUTTER MERE
Rosthwaite
Lighthouse
St. Bees Head
Ennerdale Br.
Revelin
Red Pike
High Stile
Gatescarth
Seatoller
Honister Crag
Black Lead Mine
Seathwaite
Gillerthwaite
Dent
Steeple
Pillar
Gable
Taylors Gill
Sty Head Tarn
Sprinkling Tarn
St. Bees
Copeland Forest
Egremont
Hay Cock
Kirk Fell
Wasdale H.
Great End
Pikes
Scawfell
Ehen R.
Calder R.
Greendale Tarn
Bowfell
Abbey
Calder Bridge
R. Bleng
WAST WATER
Burnmoor Tarn
Three Shire Stone
Hardknot
Cockley Beck
Druid Stones
Strands
Gosforth
WHITEHAVEN & FURNESS

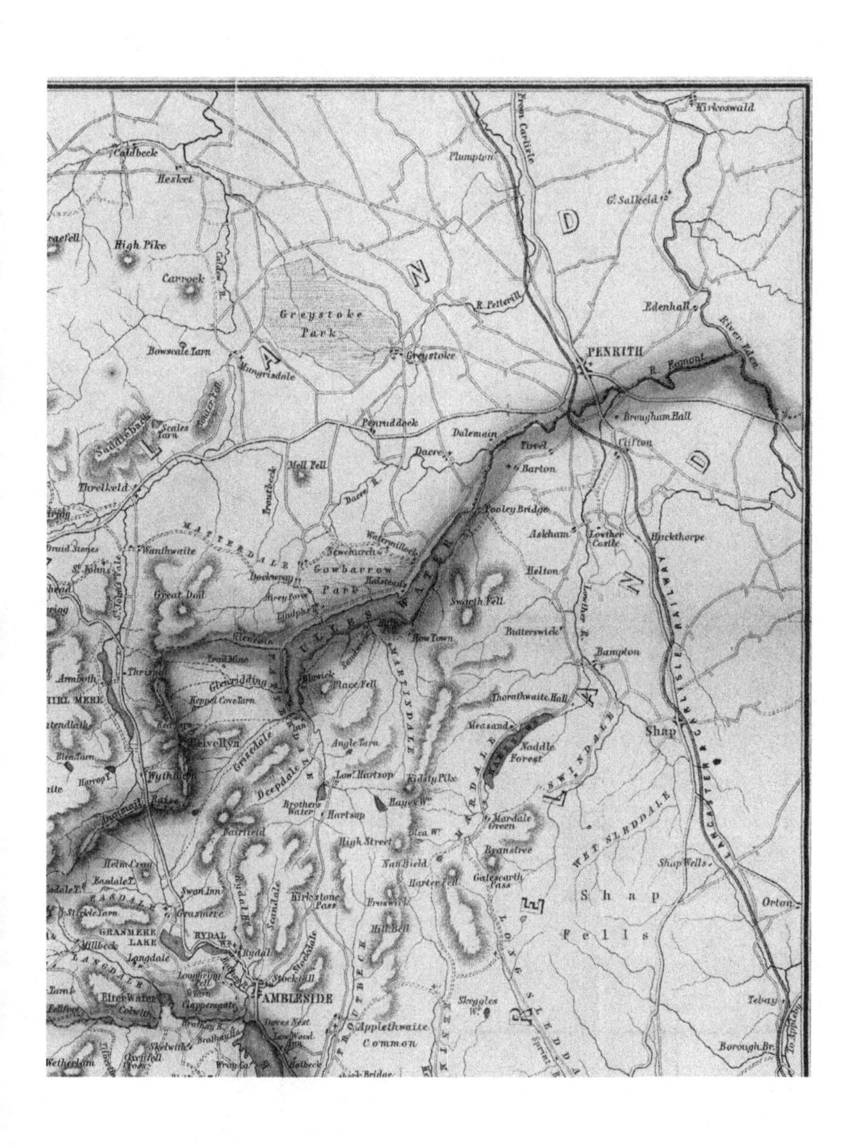

Kirkoswald
From Carlisle
Caldbeck
Plumpton
Hesket
Gt. Salkeld
High Pike
Carrock
Greystoke Park
Greystoke
R. Petteril
Edenhall
River Eden
Bowscale Tarn
PENRITH
R. Eamont
Mungrisdale
Brougham Hall
Scales Tarn
Penruddock
Dalemain
Clifton
Saddleback
Threlkeld
Mell Fell
Troutbeck
Dacre
Dacre R.
Barton
Pooley Bridge
Askham
Lowther Castle
Hackthorpe
Druid Stones
Wanthwaite
St. Johns
Gowbarrow Park
Watermillock
Newchurch
Dockwray
Halsteads
Helton
Swarth Fell
Great Dod
ULLES WATER
How Town
Butterswick
Lowther R.
Bampton
Glencoin
Lead Mine
Glenridding
Thirlspot
Armboth
THIRL MERE
Keppel Cove Tarn
Place Fell
Blowick
Thornthwaite Hall
Helvellyn
Angle Tarn
Measand
Naddle Forest
Shap
Grisedale
Deepdale
Low Hartsop
Kidsty Pike
Wythburn
Harrop T.
Blea Tarn
Hayes Wr.
Brothers Water
Hartsop
Mardale Green
High Street
Blea Wr.
Fairfield
Branstree
Nan Bield
Helm Crag
Harter Fell
Gatescarth Pass
Shap Wells
Easdale T.
Swan Inn
Kirkstone Pass
Froswick
Shap Fells
Orton
Stickle Tarn
Grasmere
GRASMERE LAKE
RYDAL
Rydal
Ill Bell
Millbeck
Langdale
AMBLESIDE
Stock Gill
Elter Water
Colwith
Clappersgate
Skeggles Wr.
Tebay
Applethwaite Common
Brathay R.
Skelwith
Doves Nest
Low Wood Inn
Borough Br.
To Appleby
Wetherlam
Oxenfell Cross
Holbeck
Wray Ca.
LANCASTER & CARLISLE RAILWAY
WET SLEDDALE
LONG SLEDDALE
SWINDALE
MARDALE
MARTINDALE
MATTERDALE
TROUTBECK
KENTMERE

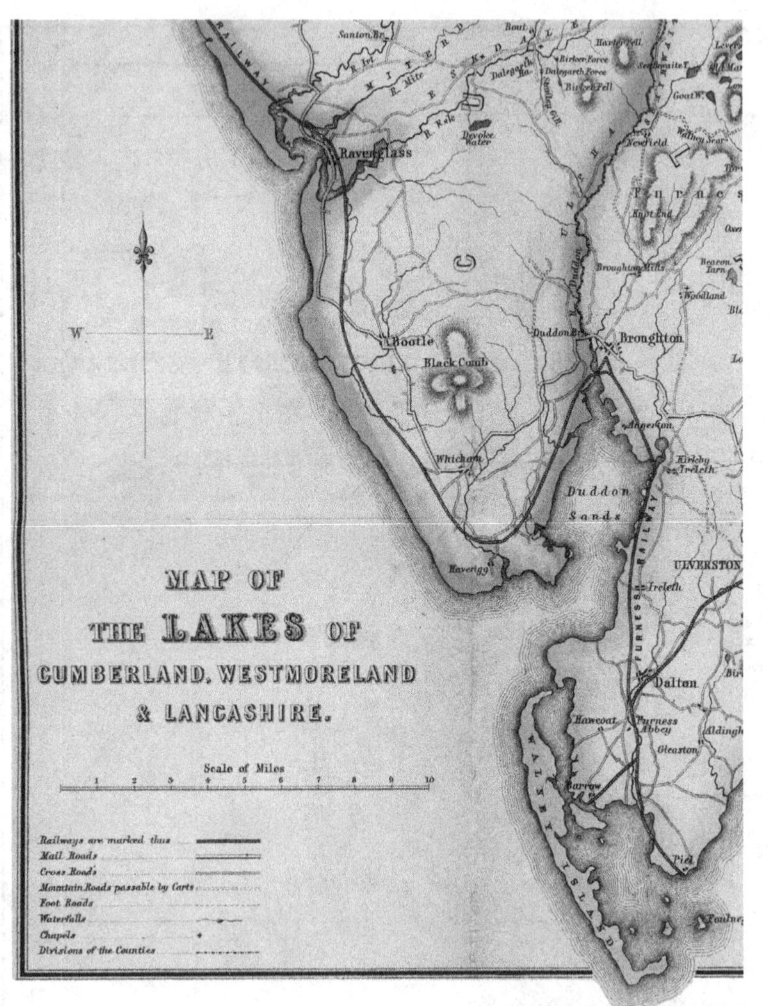

Published. by

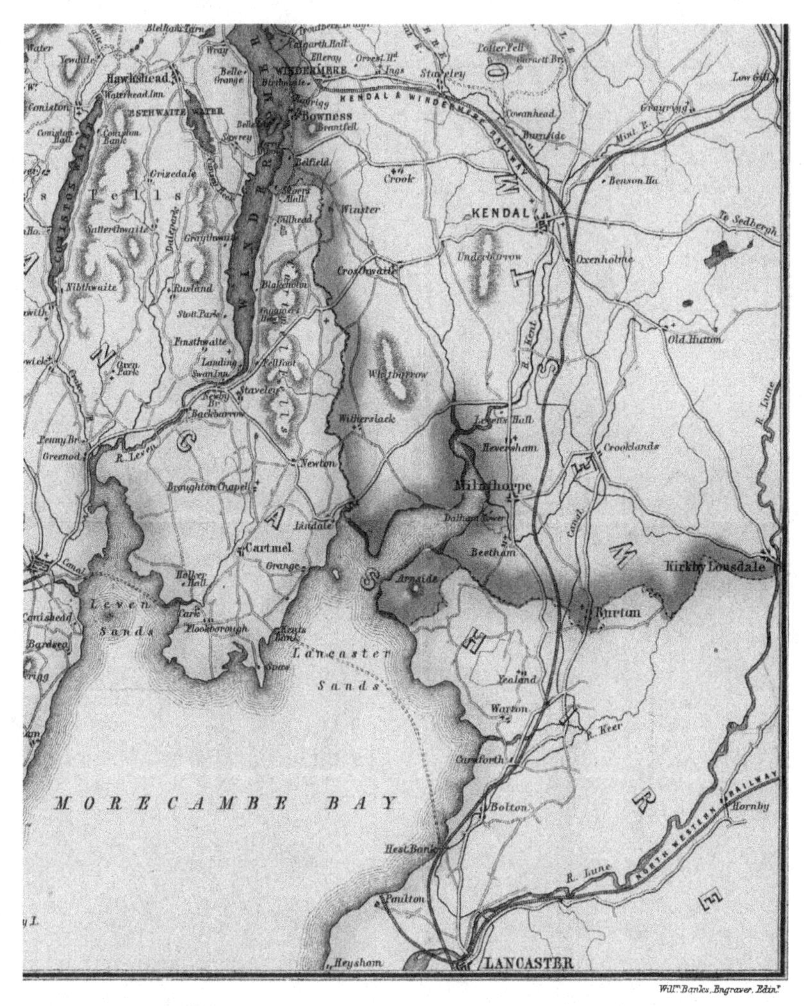

Willm. Banks, Engraver, Edinr.

John Hudson, Kendal.

Printed in the United States
By Bookmasters